# Springer Proceedings in Mathematics & Statistics

Volume 360

This book series features volumes composed of selected contributions from workshops and conferences in all areas of current research in mathematics and statistics, including operation research and optimization. In addition to an overall evaluation of the interest, scientific quality, and timeliness of each proposal at the hands of the publisher, individual contributions are all refereed to the high quality standards of leading journals in the field. Thus, this series provides the research community with well-edited, authoritative reports on developments in the most exciting areas of mathematical and statistical research today.

More information about this series at http://www.springer.com/series/10533

Xue-Cheng Tai · Suhua Wei · Haiguang Liu
Editors

# Mathematical Methods in Image Processing and Inverse Problems

## IPIP 2018, Beijing, China, April 21–24

Springer

*Editors*
Xue-Cheng Tai
Department of Mathematics
Hong Kong Baptist University
Kowloon Tong
Kowloon, Hong Kong

Haiguang Liu
Beijing Computational Science
Research Center
Beijing, China

Suhua Wei
Institute of Applied Physics
and Computational Mathematics
Beijing, China

ISSN 2194-1009 ISSN 2194-1017 (electronic)
Springer Proceedings in Mathematics & Statistics
ISBN 978-981-16-2700-2 ISBN 978-981-16-2701-9 (eBook)
https://doi.org/10.1007/978-981-16-2701-9

Mathematics Subject Classification (2010): 68U10, 94A08, 35R30, 68-XX, 35-XX



This Springer imprint is published by the registered company Springer Nature Singapore Pte Ltd.
The registered company address is: 152 Beach Road, #21-01/04 Gateway East, Singapore 189721, Singapore

Zdzislaw Trzaska

# Mathematical Modelling and Computing in Physics, Chemistry and Biology

## Fundamentals and Applications

Springer

Zdzislaw Trzaska
Warsaw University of Technology
Warsaw, Poland

ISSN 2198-4182 ISSN 2198-4190 (electronic)
Studies in Systems, Decision and Control
ISBN 978-3-031-39984-8 ISBN 978-3-031-39985-5 (eBook)
https://doi.org/10.1007/978-3-031-39985-5



This Springer imprint is published by the registered company Springer Nature Switzerland AG
The registered company address is: Gewerbestrasse 11, 6330 Cham, Switzerland

Paper in this product is recyclable.

## Program Chairs

**Xue-Cheng Tai:** General Chair, Hong Kong Baptist University
**Suhua Wei:** Organizing Chair, Institute of Applied Physics and Computational Mathematics
**Haiguang Liu:** Organizing Chair, Beijing Computational Science Research Center

## Program Committee

Raymond H. Chan, City University of Hong Kong, Hong Kong
Ke Chen, University of Liverpool, United Kingdom
Serena Morigi, University of Bologna, Italy
Michael Ng, The University of Hong Kong, China
Fiorella Sgallari, University of Bologna, Italy
Youwei Wen, Hunan Normal University, China
Haomin Zhou, Georgia University, USA

# Contents

# Point Spread Function Engineering for 3D Imaging of Space Debris Using a Continuous Exact $\ell_0$ Penalty (CEL0) Based Algorithm

**Chao Wang, Raymond H. Chan, Robert J. Plemmons, and Sudhakar Prasad**

**Abstract** We consider three-dimensional (3D) localization and imaging of space debris from only one two-dimensional (2D) snapshot image. The technique involves an optical imager that exploits off-center image rotation to encode both the lateral and depth coordinates of point sources, with the latter being encoded in the angle of rotation of the PSF. We formulate 3D localization into a large-scale sparse 3D inverse problem in discretized form. A recently developed penalty called continuous exact $\ell_0$ (CEL0) is applied in this problem for the Gaussian noise model. Numerical experiments and comparisons illustrate the efficiency of the algorithm.

**Keywords** Nonconvex optimization algorithms · 3D localization · Space debris · Point spread function

The research of the first, third and fourth authors was supported by the US Air Force Office of Scientific Research under grant FA9550-15-1-0286. The work of the second author was supported by HKRGC Grant CUHK14306316, HKRGC Grant CUHK14301718, HKRGC CRF Grant C1007-15G, HKRGC AoE Grant AoE/M-05/12, CUHK DAG No. 4053211, and CUHK FIS Grant No. 1907303. The work of the third author was also supported by HKRGC Grant CUHK14306316.

C. Wang (✉) · R. H. Chan
Department of Mathematics, The Chinese University of Hong Kong, Shatin, Hong Kong

R. J. Plemmons
Departments of Mathematics and Computer Science, Wake Forest University, Winston-Salem, NC 27109, USA

S. Prasad
Department of Physics and Astronomy, The University of New Mexico, Albuquerque, NM 87131, USA

School of Physics and Astronomy, University of Minnesota, Minneapolis, MN 55455, USA

 
X.-C. Tai et al. (eds.), *Mathematical Methods in Image Processing and Inverse Problems*, Springer Proceedings in Mathematics & Statistics 360,
https://doi.org/10.1007/978-981-16-2701-9_1

## 1 Introduction

The area of 3D imaging and localization has been getting increasing attention in recent years. The use of 3D localization in single-molecule super-resolution microscopy can obtain a complete picture of subcellubar structures [1, 6, 24]. The molecules are labeled by some specific fluorescent proteins or oligonucleotides, which can be regarded as a collection of point sources. Another application of 3D imaging is for space situational awareness (SSA). Currently, there are more than 20,000 objects in orbit around earth [27], including operational satellites, dead ones and other human-made debris. 3D localization of micro-scale space debris that become increasingly abundant with decreasing size can be vital for SSA systems responsible for the overall protection of space assets. Radar systems can sometimes detect such space debris objects, but can at best localize them with lower precision than short-wavelength optical systems. A stand-alone optical system based on the use of a light-sheet illumination and scattering concept [3] for spotting debris within meters of a spacecraft has also been proposed. A second system can localize all three coordinates of an unresolved, scattering debris [7, 25] by utilizing either the parallex between two observations, or a pulsed laser ranging system, or a hybrid system. However, to the best of our knowledge there is no other proposal of either an optical or an integrated optical-radar system to perform full 3D debris localization and tracking in the range of tens to hundreds of meters. Prasad [17] has proposed engineering point spread functions for 3D localization by the use of an optical imager that exploits off-center image rotation. This system encodes in a single image snapshot both the range $z$ and transverse $(x, y)$ coordinates of a swarm of unresolved sources such as small, sub-centimeter class space debris, which when actively illuminated can scatter a fraction of laser irradiance back into the imaging sensor. 2D image data taken with a specially designed point spread function (PSF) that encodes, via a simple rotation, changing source distance can be employed to acquire a three dimensional (3D) field of unresolved sources like space debris. Here, we propose the 3D localization and tracking of space debris at optical wavelengths by PSF engineering and employing a space-based telescope.

PSF engineering is widely used in single-molecule super-resolution [10, 15, 16, 19–21]. It is based on choosing a phase pattern that makes the defocused image of a point source depth-dependent without blurring it excessively. For space-surveillance and SSA, PSF engineering is just beginning to be considered, with [9, 18] proposing the optical theory and simulation of a single lobe rotating PSF. Such PSFs have obvious advantages over multi-lobe PSFs when dealing with high source densities at low light levels. In [26], we proposed a mathematical formulation of the 3D localization problem employing such a PSF in the Poisson-noise case. This noise model characterizes an EMCCD sensor operated in the photon-counting (PC) regime. However, when conventional CCD sensors operate at low per-pixel photon fluxes and large read-out noise, a Gaussian noise model describes more accurately this kind of data noise. In this paper, we consider the latter case.

The following forward model based on the rotating PSF image describes the spatial distribution of image brightness for $M$ point sources in the observed 2D image:

$$G(x, y) = \sum_{i=1}^{M} \mathcal{H}_{z_i}(x - x_i, y - y_i) f_i + b + \mathcal{N}(x, y), \tag{1}$$

where $\mathcal{N}$ is the Gaussian noise operator which is data-independent and $b$ is the uniform background value. Here $\mathcal{H}_{z_i}(x - x_i, y - y_i)$ is the rotating PSF for the $i$-th point source of flux $f_i$ and 3D position coordinates $(x_i, y_i, z_i)$ with the depth information $z_i$ encoded in $\mathcal{H}_{z_i}$, and $(x, y)$ is the position in the image plane. In the Fourier optics model [5] of image formation, the incoherent PSF for a clear aperture containing a phase mask that imposes an optical phase retardation, $\psi(\mathbf{s})$, on the imaging wavefront is given by

$$\mathcal{H}_z(\mathbf{s}) = \frac{1}{\pi} \left| \int P(\mathbf{u}) \exp\left[\iota(2\pi \mathbf{u} \cdot \mathbf{s} + \zeta u^2 - \psi(\mathbf{u}))\right] d\mathbf{u} \right|^2, \tag{2}$$

where $\zeta = \frac{\pi(l_0 - z)R^2}{\lambda l_0 z}$ is defocus parameter and the imaging wavelength is denoted by $\lambda$. Here $\iota = \sqrt{-1}$, $\ell_0$ is the distance between the lens and the best focus point, and $P(\mathbf{u})$ is the indicator function for the pupil of radius $R$. We use $\mathbf{s}$ with polar coordinates $(s, \phi_\mathbf{s})$ to denote a scaled version of the image-plane position vector, $\mathbf{r}$, namely $\mathbf{s} = \frac{\mathbf{r}}{\lambda z_I / R}$. Here $\mathbf{r}$ is measured from the center of the geometric (Gaussian) image point located at $\mathbf{r}_I = (x, y)$, and $z_I$ is the distance between the image plane and the lens. The pupil-plane position vector $\boldsymbol{\rho}$ is normalized by the pupil radius, $\mathbf{u} = \frac{\boldsymbol{\rho}}{R}$. For the single-lobe rotating PSF, $\psi(\mathbf{u})$ is chosen to be the spiral phase profile defined as

$$\psi(\mathbf{u}) = l\phi_\mathbf{u}, \quad \text{for} \sqrt{\frac{l-1}{L}} \le u \le \sqrt{\frac{l}{L}}, \; l = 1, \cdots, L,$$

in which $L$ is the number of concentric annular zones in the phase mask. We evaluate (2) by using the fast Fourier transform. With such spiral phase retardation, PSF (2) performs a complete rotation about the geometrical image center, as $\zeta$ changes between $-L\pi$ and $L\pi$, before it begins to degrade significantly for values of $\zeta$ outside this range.

Next, we discuss the problem of 3D localization of closely spaced point sources from simulated noisy image data obtained by using such a rotating-PSF imager. The localization problem is discretized on a cubical lattice where the coordinates and values of its nonzero entries represent the 3D locations and fluxes of the sources, respectively. Finding the locations and fluxes of a few point sources on a large lattice is evidently a large-scale sparse 3D inverse problem. Based on the Gaussian statistical noise model, we describe the results of simulation using a recently developed regularization tool called the continuous exact $\ell_0$ (CEL0) penalty term [22], which when added to a least-squares data fitting term constitutes an $\ell_0$-sparsity non-convex optimization protocol with promising results. We use an iteratively reweighted $\ell_1$ (IRL1) algorithm to solve this optimization problem.

The rest of the paper is organized as follows. In Sect. 2, we describe the CEL0-based non-convex optimization model for solving the point source localization problem. In Sect. 3, our non-convex optimization algorithm is developed. Numerical experiments, including comparisons with other optimization methods, are discussed in Sect. 4. Some concluding remarks are made in Sect. 5.

## 2 CEL0-Based Optimization Model

Here, we build a forward model for the problem based on the approach developed in [21]. In order to estimate the 3D locations of the point sources, we assume that they are distributed on a discrete lattice $\mathcal{X} \in \mathbb{R}^{m\times n\times d}$. The indices of the nonzero entries of $\mathcal{X}$ are the 3-dimensional locations of the point sources and the values at these entries correspond to the fluxes, *i.e.*, the energy emitted by the illuminated point source. The observed 2D image $G \in \mathbb{R}^{m\times n}$ can be approximated as

$$G \approx \mathcal{T}(\mathcal{A} * \mathcal{X}) + b1 + \mathcal{N},$$

where $b$ is background signal, 1 is a matrix of 1s of size the same as the size of $G$ and $\mathcal{N}$ is the Gaussian noise. Here, $\mathcal{A} * \mathcal{X}$ is the convolution of $\mathcal{X}$ with the 3D PSF $\mathcal{A}$. This 3D PSF $\mathcal{A}$ is a cube which is constructed by a sequence of images with respect to different depths of the points. Each horizontal slice is the image corresponding to a point source at the origin in the $(x, y)$ plane and at depth $z$. This tensor $\mathcal{A}$ is constructed by sampling depths at regular intervals in the range, $\zeta_i \in [-\pi L,\ \pi L]$, over which the PSF performs one complete rotation about the geometric image center before it begins to break apart. The $i$-th slice of the dictionary is $\mathcal{H}_{z_i}$ with certain depth $z_i$. Here $\mathcal{T}$ is an operator for extracting the last slice of the cube $\mathcal{A} * \mathcal{X}$ since the observed information is a snapshot.

In order to recover $\mathcal{X}$, we need to solve a large-scale sparse 3D inverse problem with data-fitting term and regularization term. Since the Gaussion noise is data-independent, it leads to the use of least squares for the data-fitting term, *i.e.*, $\frac{1}{2}\|\mathcal{T}(\mathcal{A} * \mathcal{X}) + b1 - G\|_F^2$, where $\|Y\|_F$ is the Frobenius norm of $Y$, which is equal to the $\ell_2$ norm of the vectorized $Y$. For the regularization term, we choose the continuous exact $\ell_0$ (CEL0) penalty, as described in [4, 22, 23]. It is a non-convex term approaching the $\ell_0$ norm for linear least squares data fitting problems and is constructed as

$$\mathcal{R}(\mathcal{X}) := \Phi_{\text{CEL0}}(\mathcal{X}) = \sum_{u,v,w=1}^{m,n,d} \phi(\|\mathcal{T}(\mathcal{A} * \delta_{uvw})\|_F, \mu; \mathcal{X}_{uvw}),$$

**Fig. 1** The function $\phi(a, \mu; u)$ for $\ell_2$-CEL0 with $\mu = 1$

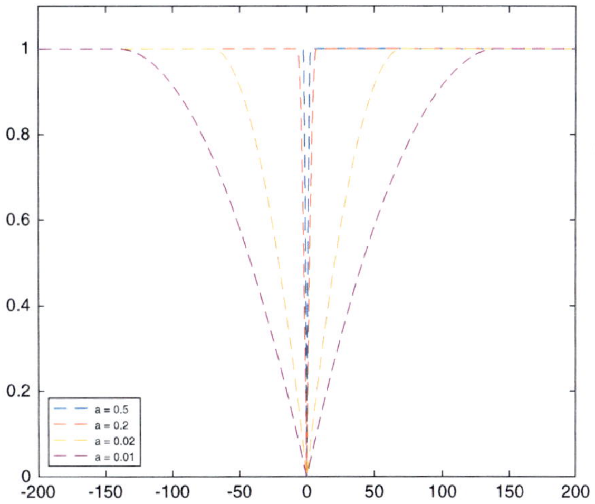


where $\phi(a, \mu; u) = \mu - \frac{a^2}{2}\left(|u| - \frac{\sqrt{2\mu}}{a}\right)^2 \mathbb{1}_{\left\{|u| \le \frac{\sqrt{2\mu}}{a}\right\}}$; see Fig. 1, and $\mathbb{1}_{\{u \in E\}} := \begin{cases} 1 & \text{if } u \in E; \\ 0 & \text{others.} \end{cases}$ Here $\delta_{uvw}$ is a 3D tensor whose only nonzero entry is at $(u, v, w)$ with value 1 and $\mu$ is the regularization parameter.

The minimization problem may be stated as

$$\min_{\mathcal{X} \ge 0} \left\{ \frac{1}{2} \|\mathcal{T}(\mathcal{A} * \mathcal{X}) + b - G\|_F^2 + \sum_{u,v,w=1}^{m,n,d} \phi(\|\mathcal{T}(\mathcal{A} * \delta_{uvw})\|_F, \mu; \mathcal{X}_{uvw}) \right\}. \quad (3)$$

To emphasize that our non-convex optimization model is based on the CEL0 regularization term, we simply designate our optimization model (3) as CEL0.

When combined with a least-squares data fitting term, CEL0 has many good properties and it does not place any strict requirements on the former. The global minimizers of the $\ell_0$ penalty model with a least squares data-fitting term ($\ell_2$-$\ell_0$) are, in fact, contained in the set of global minimizers of CEL0 (3). A minimizer of (3) can be transformed into a minimizer of $\ell_2$-$\ell_0$. Moreover, some local minimizers of $\ell_2$-$\ell_0$ are not critical points of CEL0, which means CEL0 can avoid some local minimizers of $\ell_2$-$\ell_0$.

## 3 Development of the Algorithm

Note that our optimization model for the Gaussian noise case is non-convex, due to the regularization term. We first consider an iterative reweighted $\ell_1$ algorithm (IRL1) [14] to solve the optimization problem. This is a majorization-minimization method which solves a series of convex optimization problems with a weighted-$\ell_1$ regularization term. It considers the problem (see Algorithm 3, in [14])

$$\min_{x \in X} F(x) := F_1(x) + F_2(G(x)),$$

where $X$ is the constraint set, $F$ is a lower semicontinuous (lsc) function, extended, real-valued, proper, while $F_1$ is proper, lower-semicontinous, and convex and $F_2$ is coordinatewise nondecreasing, *i.e.* $F_2(x) \leq F_2(x + te_i)$ with $x, x + te_i \in G(X)$ and $t > 0$, where $e_i$ is the $i$-th canonical basis unit vector. The function $F_2$ is concave on $G(X)$. The IRL1 iterative scheme [14, Algorithm 3] is

$$\begin{cases} W^{(k)} = \partial F_2(y), \ y = G(x^{(k)}), \\ x^{(k+1)} = \operatorname{argmin}_{x \in X} \left\{ F_1(x) + \langle W^{(k)}, G(x) \rangle \right\}, \end{cases}$$

where $\partial$ stands for subdifferential.

For the Gaussian noise problem (3), we choose

$$\begin{aligned} F_1(\mathcal{X}) &= \frac{1}{2} \|\mathcal{T}(\mathcal{A} * \mathcal{X}) + b1 - G\|_F^2 ; \\ F_2(\mathcal{X}) &= \mu - \frac{a_i^2}{2} \left( \mathcal{X}_{uvw} - \frac{\sqrt{2\mu}}{a_i} \right)^2 \mathbb{1}_{\left\{ \mathcal{X}_{uvw} \leq \frac{\sqrt{2\mu}}{a_i} \right\}} ; \\ G(\mathcal{X}) &= |\mathcal{X}| ; \\ X &= \{ \mathcal{X} \mid \mathcal{X}_{uvw} \geq 0 \text{ for all } u, v, w \}. \end{aligned}$$

Here $a_i = \|\mathcal{T}(\mathcal{A} * \delta_{uvw})\|_F$ and $i = (w-1)mn + (v-1)m + u$. The algorithm is summarized as Algorithm 1.

---

**Algorithm 1** Iterative reweighted $\ell_1$ algorithm (IRL1) for the rotating PSF problem

---

**Require:** $\mathcal{X}^{(0)} \in \mathbb{R}^{m \times n \times d}$ and $G \in \mathbb{R}^{m \times n}$. Set $\mu$.
**Ensure:** The solution $\mathcal{X}^*$ which is the minimizer in the last outer iteration.
1: **repeat**
2: Compute $W_{uvw}^{(k)} = \left( a_i \sqrt{2\mu} - a_i^2 \mathcal{X}_{uvw}^{(k)} \right) \mathbb{1}_{\left\{ \mathcal{X}_{uvw}^{(k)} \leq \frac{\sqrt{2\mu}}{a_i} \right\}}$;
3: Given $G$, $W_{uvw}^{(k)}$, obtain $\mathcal{X}^{(k)}$ by solving

$$\mathcal{X}^{(k+1)} = \operatorname*{argmin}_{\mathcal{X} \geq 0} \left\{ \frac{1}{2} \|\mathcal{T}(\mathcal{A} * \mathcal{X}) + b1 - G\|_F^2 + \sum_{u,v,w=1}^{m,n,d} W_{uvw}^{(k)} |\mathcal{X}_{uvw}| \right\}. \tag{4}$$

4: **until** convergence

---

*Remark* The minimization problem in (4) of IRL1 is a weighted $\ell_1$ model with nonnegative constraints. In [21], the $\ell_1$ model "without" nonnegative constraints is solved by the alternating direction method of multipliers (ADMM).

## 4 Numerical Results

In this section, we apply our optimization approach to solving simulated rotating PSF problems for point source localization and compare it to some other competing optimization methods. The codes of our algorithm and the others with which we compared our method were written in MATLAB 9.0 (R2016a), and all our numerical experiments were conducted on a typical personal computer with a standard CPU (Intel i7-6700, 3.4GHz).

The fidelity of localization is assessed in terms of the **recall rate**, defined as the *ratio of the number of identified true positive point sources over the number of true positive point sources*, and the **precision rate**, defined as the *ratio of the number of identified true positive point sources over the number of all point sources* obtained by the algorithm; see [1].

To distinguish true positives from false positives for the estimated point sources, we need to determine the minimum total distance between them and the true point sources. Here all 2D simulated observed images are described by 96-by-96 matrices. We set the number of zones of the spiral phase mask responsible for the rotating PSF at $L = 7$ and the aperture-plane side length as 4 which sets the pixel resolution in the 2D image (FFT) plane as 1/4 in units of $\lambda z_I/R$. The dictionary corresponding to our discretized 3D space contains 21 slices in the axial direction, with the corresponding values of the defocus parameter, $\zeta$, distributed uniformly over the range, $[-21, 21]$. According to the Abbe-Rayleigh resolution criterion, two point sources that are within $(1/2)\lambda z_I/R$ of each other and lying in the same transverse plane cannot be separated in the limit of low intensities. In view of this criterion and our choice of the aperture-plane side length and if we assume conservatively that our algorithm does not yield any significant super-resolution, we must regard two point sources that are within 2 image pixel units of each other as a single point source. Analogously, two point sources along the same line of sight (*i.e.,* with the same $x$, $y$ coordinates) that are axially separated from each other within a single unit of $\zeta$ must also be regarded as a single point source.

As in real problems, our simulation does not assume that the point sources are on the grid points. Rather, a number of point sources are randomly generated in a 3D continuous image space with certain fluxes. We consider a variety of source densities, from 5 point sources to 40 point sources in the same size space. For each case, we randomly generate 20 observed images and use them for training the parameters in our algorithm, and then test 50 simulated images with the well-selected parameters. The number of photons emitted by each point source follows a Poisson distribution with a mean of 2000 photons.

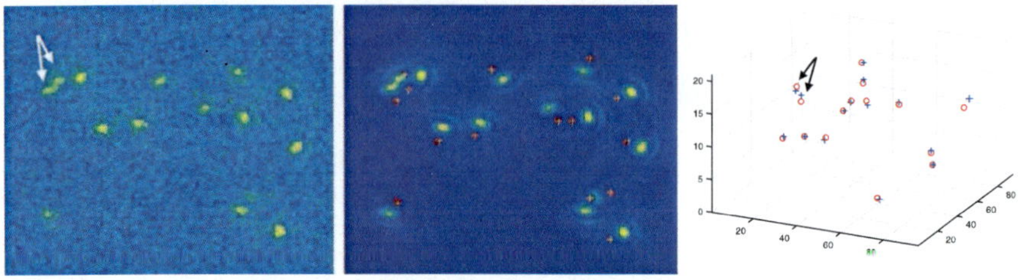

(a) Observed image (b) Estimated locations in 2D (c) Estimated locations in 3D

**Fig. 2** Localizations for the 15 point sources case. "o" denotes the location of the ground truth point source and "+" the location of the estimated point source

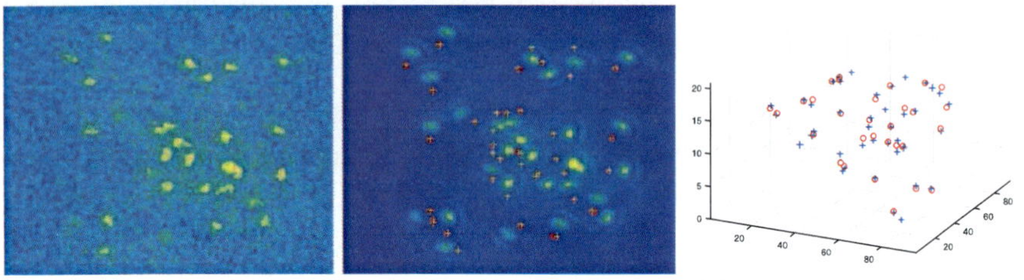

(a) Observed image (b) Estimated locations in 2D (c) Estimated locations in 3D

**Fig. 3** Localizations for the 30 point sources case

For adding the Gaussian noise, we use the MATLAB command

```
G = I0 + b + sigma*randn(Np),
```

where `b` is the uniform background noise which we set to a typical value 5. Here, `I0` is the 2D original image formed by adding all the images of the point sources without noise, and `Np` $= 96$ is the size of the images. The noise level is denoted as `sigma` and we choose it to be 10% of the highest pixel value in original image `I0`. Here, `randn` is the MATLAB command for the Gaussian distribution with the mean as 0 and standard deviation as 1.

We test our CEL0 based algorithm for several point-source densities. Figures 2 and 3 consider examples of 15 point sources and 30 point sources, respectively.

From both observed images (see Figs. 2a and 3a), neither the secondary rings nor the angle of rotation of the PSF are easily identifed, which means we cannot use a calibration method [8, 16]. In Fig. 2a, there are two overlapping PSF images corresponding to two different point sources, and our optimization approach is still able to distinguish and estimate them; see the arrows in Fig. 2a, c.

In Fig. 3, many PSF images are overlapping corresponding to point sources that are very close. Our algorithm estimates the clusters of these point sources but gives more point sources than their ground-truth number.

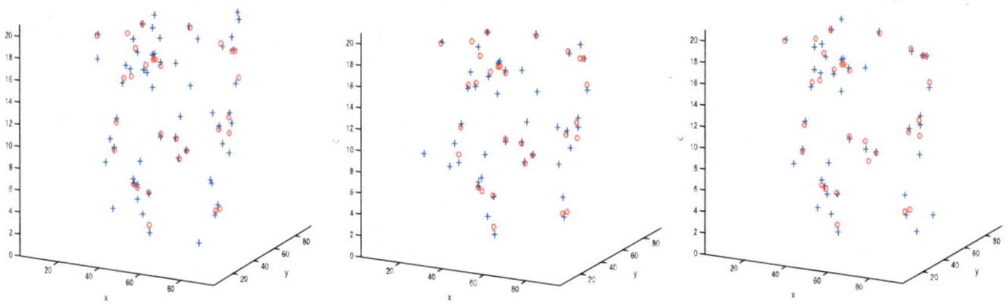

(a) $\ell_1$ (b) TL1 (c) CEL0

**Fig. 4** Localizations from three regularization models (30 point sources)

Next, we compare our algorithm with two other regularization methods: an $\ell_1$ regularization model [21] as well as a new non-convex model [12, 13, 28] called transformed $\ell_1$ (TL1). We use $\ell_1$ to denote the $\ell_1$ regularization model whose regularization term is

$$\mathcal{R}(\mathcal{X}) := \mu \|\mathcal{X}\|_1 = \mu \sum_{u,v,w=1}^{m,n,d} |\mathcal{X}_{uvw}|.$$

Following [21], we solve the optimization problem by ADMM. For TL1, the regularization term is

$$\mathcal{R}(\mathcal{X}) := \mu \sum_{u,v,w=1}^{m,n,d} \theta(a; \mathcal{X}) = \mu \sum_{u,v,w=1}^{m,n,d} \frac{|\mathcal{X}_{uvw}|}{a + |\mathcal{X}_{uvw}|},$$

where $a$ is fixed parameter and determines the degree of non-convexity. We use IRL1 to solve this model with a similar scheme as Algorithm 1. The only difference is

$$W_{uvw}^{(k)} = \partial F_2(\mathcal{X}_{uvw}^{(k)}) = \frac{a\mu}{\left(a + \mathcal{X}_{uvw}^{(k)}\right)^2},$$

where $F_2(\mathcal{Y}) = \mu \sum_{u,v,w=1}^{m,n,d} \frac{\mathcal{Y}_{uvw}}{a+\mathcal{Y}_{uvw}}$.

In Fig. 4, we again consider the 30 point sources case. We see that $\ell_2$-$\ell_1$ has more false positives than other algorithms although it detects all the ground truth point sources. TL1 and our algorithm have different but fewer false positives.

For further comparisons, we tested 50 different random images and computed the average of recall and precision rates in each density case for both algorithms; see Table 1.

In Table 1, we see that our algorithm is better than $\ell_1$ and TL1 for almost all cases especially in precision rates. For example, in the cases of 10 and 15 point sources,

**Table 1** Comparisons of $\ell_2$-$\ell_1$ with our $\ell_2$-CEL0. All the results are with post-processing

| | $\ell_1$ | | TL1 | | CEL0 | |
|---|---|---|---|---|---|---|
| No. sources | Recall (%) | Prec. (%) | Recall (%) | Prec. (%) | Recall (%) | Prec. (%) |
| 10 | 94.80 | 64.04 | 89.60 | 68.79 | 95.80 | 79.72 |
| 15 | 90.80 | 61.68 | 87.07 | 64.67 | 93.20 | 77.68 |
| 20 | 86.60 | 57.72 | 83.30 | 60.78 | 89.30 | 72.12 |
| 30 | **88.80** | 47.51 | 79.80 | 56.06 | 87.20 | 58.77 |
| 40 | **81.50** | 42.03 | 71.15 | 48.81 | 77.40 | 52.87 |

the precision rate in our algorithm is over 10% higher than the one in $\ell_1$. In the higher-density cases, like those with 30 and 40 sources, all methods have more than 5 false positives. We were able to mitigate the latter by further post-processing based on machine learning techniques, as in [21]. We set the maximum number of iterations for $\ell_1$ at 800, which guaranteed its convergence, and for CEL0 regularization and TL1 we set the maximum number of inner and outer iterations at 400 and 2, respectively. Here we emphasize the advantage of our algorithm in providing a better initial guess than $\ell_1$ and TL1 with a similar cost time.

## 5 Conclusions and Future Work

We have proposed an optimization algorithm, based on a CEL0 penalty term, for the 3D localization of a swarm of randomly spaced point sources using a rotating PSF which has a single lobe in the image of each point source. This has distinct advantages over a double-lobe rotating PSF, e.g. [11, 15, 16, 21], especially in cases where the point source density is high and the photon number per source is small. This research focuses on the Gaussian noise case which describes conventional CCD sensors in low per-pixel photon fluxes and large read-out noise. We note that at high source densities, the optimization can lead to false positives.

We employed a post-processing step based both on centroiding the locations of recovered sources that are tightly clustered and thresholding the recovered flux values to eliminate obvious false positives from our recovery sources. These techniques can be applied to other rotating PSFs as well as other depth-encoding PSFs for accurate 3D localization and flux recovery of point sources in a scene from its image data under the Poisson noise model. Applications include not only 3D localization of space debris, but also super-resolution 3D single-molecule localization microscopy, e.g. [1, 24].

Applying recently developed machine learning techniques for removing false positives instead of logistic regression models [21] will be considered. Tests of this algorithm based on real data collected using phase masks fabricated for both applications are currently being planned. In addition, work involving snapshot multispectral/hyperspectral [2] imaging, which will permit accurate material characterization, as well as higher 3D resolution and localization of space microdebris via a sequence of snapshots is under way.

## References


1. J.B. Udo, *Super-Resolution Microscopy: A Practical Guide* (Wiley, 2017)
2. H.C. Raymond, K.K. Kelvin, M. Nikolova, J.P. Robert, A two-stage method for spectral–spatial classification of hyperspectral images. J. Math Imaging Vis 1–18 (2020)
3. R.E. Christoph, J. Timothy Bays, D.M. Kenneth, M.B. Charles, C.N. Andrew, T.F. Theodore, Optical orbital debris spotter. Acta Astronautica **104**(1), 99–105 (2014)
4. S. Gazagnes, E. Soubies, L. Blanc-Féraud, High density molecule localization for super-resolution microscopy using cel0 based sparse approximation. In: 2017 IEEE 14th International Symposium on Biomedical Imaging (ISBI 2017) (IEEE, 2017)
5. W.G. Joseph, *Introduction to Fourier Optics* (Freeman, 4th edn., 2017)
6. A.-K. Gustavsson, N.P. Petar, Y.L. Maurice, Y. Shechtman, W. Moerner, 3D single-molecule super-resolution microscopy with a tilted light sheet. Nat. Commun. **9**(1), 123 (2018)
7. D. Hampf, P. Wagner, W. Riede, Optical technologies for the observation of low earth orbit objects (2015). arXiv:1501.05736
8. B. Huang, W. Wang, M. Bates, X. Zhuang, Three-dimensional super-resolution imaging by stochastic optical reconstruction microscopy. Science **319**(5864), 810–813 (2008)
9. R. Kumar, S. Prasad, PSF rotation with changing defocus and applications to 3D imaging for space situational awareness, in *Proceedings of the 2013 AMOS Technical Conference, Maui, HI* (2013)
10. D.L. Matthew, L. Steven, M. Badieirostami, W. Moerner, Corkscrew point spread function for far-field three-dimensional nanoscale localization of pointlike objects. Opt. Lett. **36**(2), 202–204 (2011)
11. E.M. William, Single-molecule spectroscopy, imaging, and photocontrol: foundations for super-resolution microscopy (nobel lecture). Angewandte Chemie Int. Edn. **54**(28), 8067–8093 (2015)
12. M. Nikolova, K.N. Michael, C.-P. Tam, Fast nonconvex nonsmooth minimization methods for image restoration and reconstruction. IEEE Trans. Image Process. **19**(12), 3073–3088 (2010)
13. M. Nikolova, K.N. Michael, S. Zhang, W.-K. Ching, Efficient reconstruction of piecewise constant images using nonsmooth nonconvex minimization. SIAM J. Imaging Sci. **1**(1), 2–25 (2008)
14. P. Ochs, A. Dosovitskiy, T. Brox, T. Pock, On iteratively reweighted algorithms for nonsmooth nonconvex optimization in computer vision. SIAM J. Imaging Sci. **8**(1), 331–372 (2015)
15. S.R.P. Pavani, R. Piestun, High-efficiency rotating point spread functions. Opt. Express **16**(5), 3484–3489 (2008)
16. P.P. Sri Rama, A.T. Michael, S.B. Julie, J.L. Samuel, L. Na, J.T. Robert, R. Piestun, W. Moerner, Three-dimensional, single-molecule fluorescence imaging beyond the diffraction limit by using a double-helix point spread function. Proc. Nat. Acad. Sci. **106**(9), 2995–2999 (2009)
17. S. Prasad, *Innovations in Space-Object Shape Recovery and 3D Space Debris Localization* (In AFOSR-SSA Workshop, Maui, 2017) Presentation slides. https://community.apan.org/wg/afosr/m/kathy/176362/download
18. S. Prasad, Rotating point spread function via pupil-phase engineering. Opt. Lett. **38**(4), 585–587 (2013)
19. Y. Shechtman, J.S. Steffen, S.B. Adam, W. Moerner, Optimal point spread function design for 3d imaging. Phys. Rev. Lett. **113**(13), 133902 (2014)

20. Y. Shechtman, E.W. Lucien, S.B. Adam, J.S. Steffen, W. Moerner, Precise three-dimensional scan-free multiple-particle tracking over large axial ranges with tetrapod point spread functions. Nano Lett. **15**(6), 4194–4199 (2015)
21. S. Bo, W. Wang, H. Shen, J.T. Lawrence, C. Flatebo, J. Chen, A.M. Nicholas, D.C.B. Logan, F.K. Kevin, F.L. Christy, Generalized recovery algorithm for 3d super-resolution microscopy using rotating point spread functions. Sci. Rep. **6** (2016)
22. E. Soubies, L. Blanc-Féraud, G. Aubert, A continuous exact $\ell_0$ penalty (cel0) for least squares regularized problem. SIAM J. Imaging Sci. **8**(3), 1607–1639 (2015)
23. E. Soubies, L. Blanc-Féraud, G. Aubert, A unified view of exact continuous penalties for $\ell_2$-$\ell_0$ minimization. SIAM J. Optim. **27**(3), 2034–2060 (2017)
24. A. von Diezmann, Y. Shechtman, W. Moerner, Three-dimensional localization of single molecules for super-resolution imaging and single-particle tracking. Chem. Rev. (2017)
25. P. Wagner, D. Hampf, F. Sproll, T. Hasenohr, L. Humbert, J. Rodmann, W. Riede, Detection and laser ranging of orbital objects using optical methods, in *Remote Sensing System Engineering VI*, vol. 9977 (International Society for Optics and Photonics, 2016), p. 99770D
26. C. Wang, R. Chan, M. Nikolova, R. Plemmons, S. Prasad, Nonconvex optimization for 3-dimensional point source localization using a rotating point spread function. SIAM J. Imaging Sci. **12**(1), 259–286 (2019)
27. A. Witze, The quest to conquer earth's space junk problem. Nature **561**, 24–26 (2018)
28. S. Zhang, J. Xin, Minimization of transformed $\ell_1$ penalty: closed form representation and iterative thresholding algorithms. Commun. Math. Sci. **15**(2), 511–537 (2018)

# An Adjoint State Method for An Schrödinger Inverse Problem

**Siyang Wei and Shingyu Leung**

**Abstract** We propose a simple algorithm for solving an inverse problem for the Schrödinger equation. The idea is to apply the gradient descent and the adjoint state technique. We observe that since the forward operator is self-adjoint, the approach simply requires to solve the same partial differential equation for both the forward problem and the adjoint problem. To speed up the computations, we also develop a cascadic initialization strategy to provide a better initial condition for the inversion process. To be more realistic for real life applications, we incorporate techniques from the level set method to handle cases with only a set of finite number of Dirichlet-to-Neumann (DN) measurements. Moreover, based on a usual reduction, this inverse problem can be linked to the standard Calderón inverse problem for the electrical impedance tomography (EIT). Therefore, our approach might provide a simple numerical alternative to solve the EIT problem. Numerical examples will demonstrate that the new formulation is effective and robust.

**Keywords** Inverse problem · Adjoint state method · Schrödinger equation · EIT

## 1 Introduction

We consider the following Schrödinger equation

$$(-\Delta + q)\,u = 0 \ \text{ in } \ \Omega\,, \tag{1.1}$$

S. Wei
Department of Mathematics, University of California at Irvine, Irvine, CA 92697-3875, USA
e-mail: siyangw2@uci.edu

S. Leung (✉)
Department of Mathematics, The Hong Kong University of Science and Technology, Clear Water Bay, Hong Kong
e-mail: masyleung@ust.hk



X.-C. Tai et al. (eds.), *Mathematical Methods in Image Processing and Inverse Problems*, Springer Proceedings in Mathematics & Statistics 360,
https://doi.org/10.1007/978-981-16-2701-9_2

where $\Omega \subset \mathbb{R}^d$ is an open bounded domain and $q \in L^\infty(\Omega)$ is a potential. Assuming that zero is not a Dirichlet eigenvalue, we define the Dirichlet-to-Neumann (DN) map by

$$\Lambda_q : u|_{\partial\Omega} \mapsto \partial_{\mathbf{n}} u|_{\partial\Omega}$$

where $\mathbf{n}$ is the unit outward normal to $\partial\Omega$ on the boundary, which will be denoted as an operator $\Lambda_q : H^{1/2}(\partial\Omega) \to H^{-1/2}(\partial\Omega)$. In this paper, we are interested in developing a numerical method for solving the corresponding inverse problem, i.e. given the DN map on the boundary of the domain, we are going to invert for the potential $q$ in $\Omega$.

This inverse problem has a wide range of applications including reflection seismology from geosciences assuming a time-harmonic wave description. It also has a close link to other elliptic inverse problems such as the electrical impedance tomography (EIT) [35]. EIT is a noninvasive type of medical imaging, where currents are applied to the surface of the body $\partial\Omega$ and the induced voltage is used to reconstruct the interior electrical conductivity and permittivity. The mathematical model of this imaging technique can be described as follows. We denote the bounded electrical conductivity inside $\Omega$ by $\gamma(\mathbf{x}) > 0$. The electric potential or voltage is denoted by $v(\mathbf{x}) \in H^1(\Omega)$. Then $v(\mathbf{x})$ satisfies the anisotropic conductivity equation

$$\nabla \cdot \gamma \nabla v = 0 \tag{1.2}$$

with the Dirichlet boundary condition $v|_{\partial\Omega} = f$. The DN map, or the voltage to current map in the physical application, is given by $\Lambda_\gamma(f) = \gamma \partial_{\mathbf{n}} v|_{\partial\Omega}$, where $\mathbf{n}$ denotes the unit outward normal to $\partial\Omega$. The inverse conductivity problem is to recover the conductivity function $\gamma(\mathbf{x})$ inside the domain knowing $\Lambda_\gamma$. This problem was first proposed in [8] and is commonly known as the Calderón inverse problem.

It is possible to reformulate this Calderón inverse problem into the inverse problem we are concentrating in this work. If we define a function $q \in L^\infty(\Omega)$ satisfying

$$(-\Delta + q)\sqrt{\gamma} = 0 \tag{1.3}$$

and introduce $u = \sqrt{\gamma}\, v$, we can then easily transform (1.2) into the elliptic equation (1.1) with the corresponding DN map. In the past few decades, various numerical approaches have been developed to solve this ill-posed Calderón EIT inverse problem. These numerical methods can be roughly categorized into two classes. The first class is a direct method which tries to explicitly construct the conductivity function. A construction algorithm based on the Born series has been developed in [2]. A bayesian inversion algorithm has been developed in [21], while a factorization method is proposed in [7]. The second class is iterative methods which tries to invert the conductivity through optimization procedure. For example, the NOSER algorithm developed in [12] takes one step of a Newton's method with constant conductivity as initial guess. Later, the CNRSER algorithm [14] takes multiple Newton-Raphson steps and iteratively solves the inverse admittivity problem. Another Newton's based approach can also be found in [23].

Related to the EIT, the level set method has also been incorporated into the optimization procedure to recover piecewise constant conductivity functions [9, 13]. Some other optimization approaches to the piecewise constant case can be found in [4, 10, 22]. For partial measurements, [18] has recently proposed a direct inversion approach based on the so-called D-bar method [17, 29, 30].

There are some theoretical results regarding the uniqueness of the solution $\gamma(\mathbf{x})$. An early work in [34] showed that the DN map $\Lambda_\gamma$ uniquely determines the conductivity in dimension $n \geq 3$. For the two dimensional case, [30] proved that a $C^2$ conductivity function can be uniquely reconstructed with respect to the corresponding DN map. More recently, the uniqueness result has been extended to Lipschitz conductivities in [6]. Clearly we are not able to prove a complete summary of this research field here. Instead, we refer all interested readers to some survey and review papers such as [5, 11, 36, 37].

Except through the change of variable to the EIT problem, there are also stability and uniqueness results directly for the inverse problem for the Schrödinger equation. Some stability results were first developed in [28]. A Lipschitz-type stability result has recently been established in [3] which assumes a priori that the potential is piecewise constant with a bounded known number of unknown values. For the case with finite number of boundary measurements, it has been recently proven in [1, 15] that one can still uniquely determine the $L^\infty$ potential in the Schrödinger equation.

Even with such a tight connection to the EIT problem, we note that there are very few numerical procedure to the inverse problem for the Schrödinger equation. In this paper, we propose an adjoint state method directly to this inverse problem. Because the forward operator is self-adjoint, the overall algorithm requires to solve the same type of equations for all the forward problem, the adjoint problem and the regularization problem. Since this key equation is simply a standard linear elliptic partial differential equation, there are well-developed numerical solvers which, therefore, make the implementation of the solution procedure very straight-forward.

Moreover, our algorithm can also be regarded as a simple alternative for EIT. Numerically, it is hard to maintain the positivity of the conductivity function $\gamma$ when solving the EIT problem using an iterative approach. Especially in a gradient descent approach, one has to pick a time step small enough so that the update would not create a negative conductivity. Otherwise the forward problem itself will then be ill-posed and, therefore, the numerical procedure will be crashed. Our approach, on the other hand, has no restriction on the positivity of the potential $q$. Constructing the solution to the EIT from that of the inverse problem for the Schrödinger, we could simple solve (1.3) which is essentially the same PDE for all key steps in our algorithm.

We will first explain our adjoint state approach to recover the Schrödinger potential $q$ from the DN map in Sect. 2. In Sect. 3, we provide some implementation details about the proposed algorithm and will discuss two generalizations of the method. In the derivation, we assume that measurements can be given on the whole computational domain. In this section, we first incorporate some common techniques as in the level set method [31–33] to handle cases with finite number of measurements on a set of locations on the boundary. Then we will develop a cascadic approach to improve the overall computational efficiency by providing a better initial condition obtained from a coarse mesh. We will give some numerical tests to demonstrate the effectiveness and the robustness of our proposed algorithm in Sect. 4. Finally, we conclude the paper in Sect. 5 with some possible future directions.

## 2 Our Proposed Approach Based on the Adjoint State Method

In this work we are interested in the Schrödinger inverse problem: we would like to invert for the potential $q(\mathbf{x})$ inside the domain $\Omega$ from given a finite number of measurements $\partial_{\mathbf{n}}u$ on the boundary $\partial\Omega$. To achieve this, we propose to invert for the potential $q$ by a variational approach. The simplest model is to minimize the the $L^2$-difference between the experimental measurements and those obtained from solving the Schrödinger equation on the boundary of the computational domain. Mathematically, we propose the following mismatching energy designed base on the least squares idea,

$$E(q) = \frac{1}{2}\int_{\partial\Omega} \left|\partial_{\mathbf{n}}u - \partial_{\mathbf{n}}u^*\right|^2 , \tag{2.1}$$

where $\partial_{\mathbf{n}}u^*|_{\partial\Omega}$ is the given measurement and $\partial_{\mathbf{n}}u|_{\partial\Omega}$ is computed by solving (1.1) together with the boundary condition

$$u|_{\partial\Omega} = u^*|_{\partial\Omega} . \tag{2.2}$$

To minimize the mismatching energy, we compute the Euler-Lagrange equation and apply the method of gradient descent. We first perturb the potential $q$ by $\epsilon\tilde{q}$, which induces a corresponding change in $u$ by $\epsilon\tilde{u}$. The change in the overall energy is then given by

$$\delta E = \epsilon \int_{\partial\Omega} \partial_{\mathbf{n}}\tilde{u}\left(\partial_{\mathbf{n}}u - \partial_{\mathbf{n}}u^*\right) + O(\epsilon^2) . \tag{2.3}$$

Now, from the state Eq. (1.1), the perturbations in both functions $q$ and $u$ are related by the elliptic equation

$$(-\Delta + q)\tilde{u} = -\tilde{q}u . \tag{2.4}$$

We need to determine the perturbation in the potential $q$, denoted by $\tilde{q}$, so as to decrease the energy $\delta E$. The main difficulty is that the perturbation in $E$, denoted by $\delta E$, depends implicitly on $\tilde{q}$ through $\tilde{u}$ and the partial differential equation (2.4).

To efficiently compute $\tilde{q}$ which minimizes $E$, we apply the adjoint state method. Multiplying (2.4) by an adjoint variable $\lambda$, integrating it over $\Omega$, applying integration by parts, and adding the resulting expression to (2.3), we finally have

$$\frac{\delta E}{\epsilon} = \int\limits_{\partial\Omega} \partial_{\mathbf{n}}\tilde{u}\left(\partial_{\mathbf{n}}u - \partial_{\mathbf{n}}u^*\right) + \int\limits_{\partial\Omega} \lambda\partial_{\mathbf{n}}\tilde{u} + \int\limits_{\Omega} \tilde{u}(\Delta - q)\lambda - \int\limits_{\Omega} \lambda\tilde{q}u$$
$$- \left( \int\limits_{y}\int\limits_{z} \lambda_x \tilde{u}\big|_{x_{\min}}^{x_{\max}} dydz + \int\limits_{x}\int\limits_{z} \lambda_y \tilde{u}\big|_{y_{\min}}^{y_{\max}} dxdz + \int\limits_{x}\int\limits_{y} \lambda_z \tilde{u}\big|_{z_{\min}}^{z_{\max}} dxdy \right) + O(\epsilon)\,.$$

Now, if we are able to choose $\lambda$ satisfying

$$(-\Delta + q)\lambda = 0\,, \tag{2.5}$$

with the boundary condition,

$$\lambda|_{\partial\Omega} = \left(\partial_{\mathbf{n}}u^* - \partial_{\mathbf{n}}u\right)\big|_{\partial\Omega} \tag{2.6}$$

on the boundary $\partial\Omega$, one can eliminate the dependence of $\tilde{u}$ when determining the gradient of $E$ with respect to $q$. We call Eq. (2.5) the adjoint state equation. We note that this adjoint state equation is actually the same as the state Eq. (1.1) since the operator is in fact self-adjoint.

Ignoring all higher order terms in the energy perturbation, we have

$$\frac{\delta E}{\epsilon} = \int\limits_{\Omega} -\lambda\tilde{q}u\,.$$

To minimize the energy using the method of gradient descent, one could choose the perturbation $\tilde{q} = \lambda u$. This implies that $\delta E = -\epsilon \int\limits_{\Omega} \lambda^2 u^2 \leq 0$ and the equality holds when $\|\lambda u\|_{H^{1/2}(\Omega)} = 0$. However, such choice of the gradient direction does not necessary guarantee the smoothness in the inverted potential $q$.

To have a stable algorithm, we require a regularity condition on $q^k$ that it should be a smooth function. This regularity seems to be too restrictive in practice. In general, one only needs $q^k \in C^1$ to guarantee well-posedness of the state Eq. (1.1). However, assuming that we assign $\tilde{q} = \lambda u$ directly, it is not clear whether this function would give us the desired regularity. Even if this perturbation is in $C^1$, the numerical solution may have jumps or spikes. These irregularities will force one to pick a very small step-size, $\epsilon$, in the minimization process. Therefore, to have a faster convergence, we impose the following elliptic regularity in each iteration. In this work, we propose to use the descent direction

$$\tilde{q} = (I - \nu\Delta)^{-1}(\lambda u)\,, \tag{2.7}$$

with the homogeneous Dirichlet boundary condition

$$\tilde{q}|_{\partial\Omega} = 0 \tag{2.8}$$

where $I$ is the identity operator and $\nu \geq 0$ controls the amount of regularity we want. If we take a larger $\nu$, the resulting $\tilde{q}$ would be smoother. The boundary condition assumes that we can measure $q^*|_{\partial\Omega}$, which is reasonable. With this particular $\tilde{q}$, we have

$$\delta E = -\epsilon \int_{\Omega} (\tilde{q}^2 - \nu \tilde{q} \Delta \tilde{q}) = -\epsilon \int_{\Omega} (\tilde{q}^2 + \nu \|\nabla \tilde{q}\|^2) \leq 0 \,.$$

To start the gradient descent, we need to initialize the iteration by assigning a potential function $q^0$. In this work, we propose to determine the initial condition by solving the following elliptic equation

$$(-\Delta + \alpha)\, q^0 = 0 \tag{2.9}$$

with the boundary condition $q^0|_{\partial\Omega} = q^*|_{\partial\Omega}$ and the constant parameter $\alpha \geq 0$ controlling the regularity in the initial condition. In general, we recommend simply solving the standard Laplace equation (i.e. with $\alpha = 0$) to initial the iteration in practice. For the special case when the exact potential $q^*$ is in fact a constant function, however, we note that such an initialization process will lead to the exact recovery. Therefore, in the following numerical examples, we choose a nonzero $\alpha$ for the constant test case in Sect. 4.1 but $\alpha = 0$ for other cases.

# 3 Some Implementation Details

In this section, we first summarize the above algorithm and provide some implementation details. Even though the above adjoint state method seems to be straightforward, it requires full measurements on the computational boundary. In this section, we will also generalize the approach to handle the case with only a set of finite number of measurements on the computational boundary. In principle, the generalization can be applied to the case when the measurements are located in the interior of the domain, but the idea is similar to what we are going to discuss and it will be omitted here. Finally, because the approach based on the gradient descent converges exponentially, it will take a large number of iterations until the steady state solution. We are also going to propose a cascadic initialization strategy which seems to be able to significantly improve the computational efficiency.

## *3.1 Inversion with Full Measurements*

Here we give an algorithm for this Schrödinger inverse problem with full measurements of $\partial_{\mathbf{n}} u$ on the boundary.

In the above algorithm, Eqs. (1.1), (2.5) and (2.7) all require to invert the same linear elliptic operator. Since it is a rather straightforward and standard numerical procedure to obtain a numerical approximation to the equation, we will omit the details in the numerical procedure in this article. The iteration size $\epsilon^k$ can be chosen

---

**Algorithm 1** Full measurements on the computational boundary.

1: Discretize the computational domain using a $(N+1) \times (N+1)$ mesh. Set $k = 0$ and initialize $q^0$ by solving (2.9).
2: **while** $\|\tilde{q}^k\|_2 \leq \delta$ or $k \geq k_{\max}$ for some constants $\delta$ and $k_{\max}$ **do**
3: $\quad k = k + 1$
4: $\quad$ Obtain $u^k$ by solving (1.1) with the boundary condition (2.2) using $q = q^k$
5: $\quad$ Obtain $\lambda^k$ by solving (2.5) with the boundary condition (2.6)
6: $\quad$ Obtain $\tilde{q}^k$ using (2.7) with the boundary condition (2.8)
7: $\quad$ Update $q^{k+1} = q^k + \epsilon^k \tilde{q}^k$
8: **end while**

---

to be a small enough constant for simplicity or based on some carefully designed optimization procedure such as the Armijo-Goldstein rule in the typical line search method.

Finally, we comment that it is possible to further improve the computational efficient by replacing the standard gradient descent approach here by a derivative-based approach as in [24]. In particular, one can apply the quasi-Newton method defined by $q^{k+1} = q^k + \epsilon^k s^k$ where $s^k = -A_k^{-1} E'(q^k)$ and $A_k$ is a positive definite operator satisfying the secant condition

$$A_{k+1}(q^{k+1} - q^k) = E'(q^{k+1}) - E'(q^k) .$$

In the iteration, the operator $A_{k+1}$ is updated by modifying the previous operator $A_k$ using procedure like the Broydon-Fletcher-Goldfarb-Shanno (BFGS) method. But we are not going to further investigate this approach in the article but will leave it as a future work.

## 3.2 *Inversion with Finite Number of Measurements*

In the above discussion, we have assumed that the DN map is given on the whole computational boundary. In this section, we propose a simple numerical procedure to relax this assumption so that the inverse problem is more realistic. Mathematically, we let $\Gamma \subset \partial\Omega$ be the set of locations where we have measurements. The formulation we are going to propose allows that it be a segment of the boundary or even be more flexible so that it is just a finite number of sampling locations.

Now, since we have measurements only on $\Gamma$, we replace the original mismatch functional (2.1) by the following

$$E(q) = \frac{1}{2} \int_{\Gamma} \left|\partial_{\mathbf{n}} u - \partial_{\mathbf{n}} u^*\right|^2 = \frac{1}{2} \int_{\partial\Omega} \left|\partial_{\mathbf{n}} u - \partial_{\mathbf{n}} u^*\right|^2 \delta(\Gamma)$$

where $\delta(x)$ is the Dirac's delta function. Such modification has no effect in the forward problem. The adjoint equation itself is still the same as in Eq. (2.5) while the corresponding boundary condition (2.6), however, is modified to incorporate the measurements set $\Gamma$ into the formulation,

$$\lambda|_{\partial\Omega} = \left(\partial_{\mathbf{n}} u^* - \partial_{\mathbf{n}} u\right) \delta(\Gamma) . \tag{3.1}$$

Once we modify this boundary condition to the adjoint equation, the whole algorithm follows and so we are not presenting the overall algorithm here.

Numerically, we follow the standard approach from the level set method [31–33] and approximate the delta function based on the smoothed Heaviside step function as follows:

$$H(x) := \begin{cases} 0, & x < -\eta \\ \frac{1}{2} + \frac{x}{2} + \frac{\eta}{2\pi}\sin(\frac{\pi x}{\eta}), & -\eta < x < \eta \\ 1, & x > \eta, \end{cases}$$

$$\delta(x) := H'(x) = \frac{1}{4}\left[1 - \text{sign}(|x| - \eta)\right]\left[1 + \cos\left(\frac{\pi x}{\eta}\right)\right], \tag{3.2}$$

where $\eta > 0$ is a parameter controlling the width of the smoothing window. In the usual level set method, this parameter $\eta$ is recommended to be proportional to the computational mesh size and is given by $\eta = 2.5\Delta x$.

Now we consider the boundary potential $q^*$. In applications where one assumes the far field potential is a constant or is well-approximated by a constant, one might assume that these measurements are available on the boundary of the computational domain so that all previous procedure related to the potential $q$ follows. In some other applications where the measurements of the DN map are sampled uniformly along the computational boundary given by $\Gamma$, we assume that the potential measurements $q^*$ are provided at the same set of finite locations. To start the algorithm, we approximate the measurements on $\partial\Omega$ by interpolation and treat the interpolant on the whole computational boundary as the given measurements. Then the original dicussion follows.

**Algorithm 2** An inversion algorithm based on the cascadic initialization strategy

1: Pick two constants $M_c < M_f$ such that the mesh on the coarsest and the finest levels are given by $\left(2^{M_c}+1\right) \times \left(2^{M_c}+1\right)$ and $\left(2^{M_f}+1\right) \times \left(2^{M_f}+1\right)$, respectively.
2: Follow Algorithm 1 using a mesh with $N = 2^{M_c}$.
3: **while** $N$ is not $2^{M_f}$ **do**
4: Set $N = 2N$.
5: Interpolate the latest solution $q^\infty$ onto the finer mesh and using it as the initial condition for Algorithm 1.
6: **end while**

### 3.3 A Cascadic Initialization Approach

In the above algorithms, we have proposed to initialize the iteration by the solution to a simple elliptic equation (2.9). Such choice unfortunately might not give a good initial guess for the iteration in practice and the resulting algorithm might perform inefficiently. It will need a less number of time iterations if one is able to provide an initial condition closer to the exact solution. To improve the overall computational efficiency, we propose the following cascadic initialization approach to obtain a better initial guess for the gradient descent. Such cascadic strategy has been shown to be very effective in other applications including the computations of the effective Hamilonian for a class of Hamilton-Jacobi operators [16].

We first invert for the potential on a relatively coarse mesh using the initial condition proposed above. As discussed above, this initial condition could be very different from the exact solution and could take a large number of iterations until it reaches the converged solution. However, the computational cost on a coarse mesh is still cheap. Now, since the solution to the problem is supposed to be $L^\infty$, we propose to interpolate this coarse mesh solution onto a finer grid and use the interpolant as the initial condition on the finer grid. To simplify the implementation of the algorithm, we can double the number of mesh points in each physical dimension by inserting an extra sampling point in the middle of any two adjacent grid points on the coarse level. Then this approach can reuse all measurements in the coarse level, yet seems to provide a good initial condition for the computations on the fine level. Once we have the converged inversion on this new level, we can repeat the whole process and further refine the computational mesh. The overall algorithm is summarized in Algorithm 2.

## 4 Numerical Examples

In the following examples, the finest mesh is given by $161 \times 161$ grid points in the $x - y$ plane. For each of the Schrödinger model below, we have implemented one case with full measurements on the whole boundary of the computational domain $\partial_{\mathbf{n}} u$, and another case with finite number of measurements on the boundary. In the case with finite number of measurements of $\partial_{\mathbf{n}} u|_{\partial\Omega}$, we have chosen sampling points uniformly on the boundary $\partial\Omega$ and have applied the delta function (3.2) with $\eta = 2.5\Delta x$, which depends on the mesh size in each level of the cascadic approach. To speed up the implementation, we are going to also test the cascadic algorithm for the following cases. To initialize the whole process, we assign the initial condition $q^0$ by solving the elliptic equation (2.9) with the boundary condition $q^0|_{\partial\Omega} = q^*$ with $\alpha = 0$ for most cases except the constant case where we pick $\alpha = 1$. We perform the cascadic approach with 5 levels in total, where the mesh on the coarsest level and the finest level are given by $11\times 11$ and $161\times 161$, respectively.

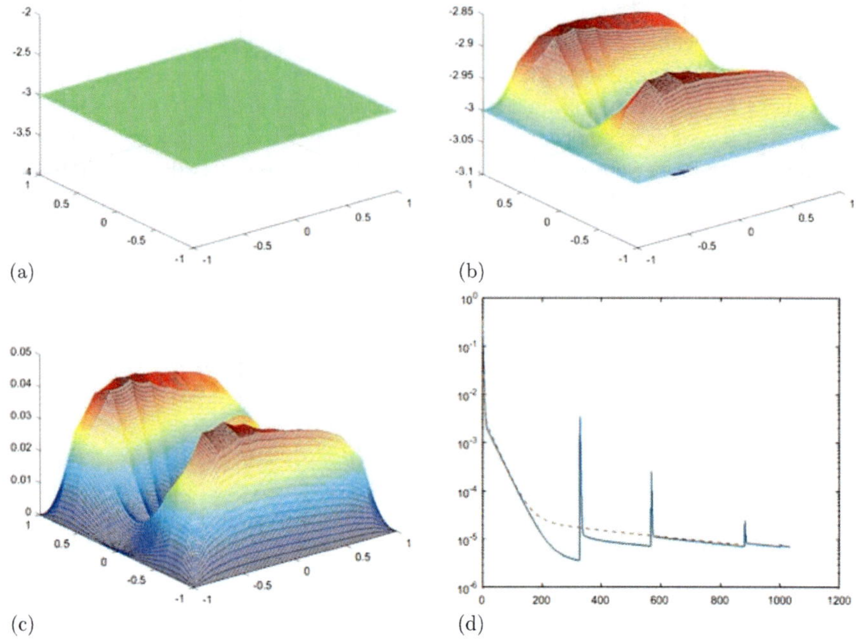


**Fig. 1** (Section 4.1) **a** The exact constant $q$, **b** the inverted potential $q$ from full measurements and **c** the relative error, and **d** the convergence history (Blue solid line: the cascadic approach. Red dashed line: the direct approach)

## 4.1 A Constant Model

In the first example, we consider the following model with the exact solution given by $u = e^x \cos 2y$ and a constant potential $q \equiv -3$ with the computational domain given by $[-1, 1]^2$. We terminate the iterations while $\left|E^k - E^{k-1}\right| \leq 10^{-8}$. The numerical results are shown in Figs. 1 and 2. Figure 1 shows the numerical solution proposed by our adjoint state method. When full measurements are given on the whole computational domain, we found that our solution has 4.03% error in the inverted solution comparing to the exact solution. There are spikes in the convergence history in Fig. 1d because the cascadic initialization approach refines the mesh and this increases the number of measurements from the boundary points. Therefore, the mismatch in the DN map is significantly increased. As the refinement goes on, the magnitude of the spike decreases since the numerical solution gets closer to the exact solution which gives a more accurate DN map.

To see how the cascadic initialization improves the computational efficiency, we directly obtain the solution on the finest mesh $161\times161$. The corresponding convergence history is plotted on top of Fig. 1d using a red dashed line. Although the total

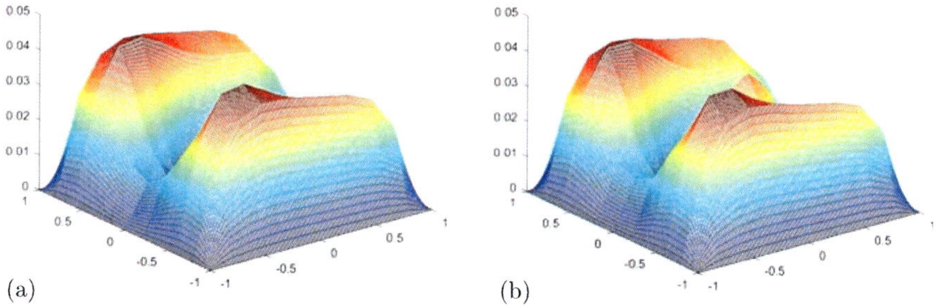


**Fig. 2** (Section 4.1) The relative error in the inverted potential $q$ with **a** 80 and **b** 20 measurements.eps

number of iterations in the cascadic approach slightly increases, the overall computational time is still greatly improved since the cascadic initialization approach requires computations mostly on the coarse meshes.

In Fig. 2, we have shown the inverted solutions corresponding to the case when we only have finite number of measurements on the boundary. In Fig. 2a, we assume that the boundary potential $q^*$ is known on the whole computational boundary and invert the interior potential $q$ using only 80 DN map measurements with the sampling locations uniformly placed on the boundary. The relative error in the solution is now increased to 4.26%. Even thought the relative error is further increased when we keep reducing the number of measurements to 20, as shown in Fig. 2b, the percentage error in the solution is still below 4.32% and is still acceptable.

## 4.2 A Non-constant Model

In this example, we consider another smooth solution given by $u = \exp(\sin x + \cos y)$ on the domain $[-1, 1]^2$ with a non-constant potential $q = \cos^2 x - \sin x + \sin^2 y - \cos y$. Note that this potential function $q$ changes sign within the computational domain. Here, we terminate the iterations when $\left|E^k - E^{k-1}\right| \leq 10^{-4}$ is reached which is larger than the tolerance we use in the previous example. Even though the choice of such parameter is indeed in general problem-dependent, we find that the quality of the inverted solution does not significantly depend on this tolerance.

Figures 3, 4 and 5 show our computed solutions. Figure 3 shows the solution obtained with full (640) measurements of both DN map and the Schrödinger potential $q$ on the boundary. We can see that the inverted $q$ is almost the same as the exact $q$. To better see the convergence in the numerical approach, we plot only the absolute error in both the initial condition of the potential $q$ and also the final solution. The largest absolute error in the domain is dropped from around 0.4 in the initial guess to approximately 0.08 eventually.

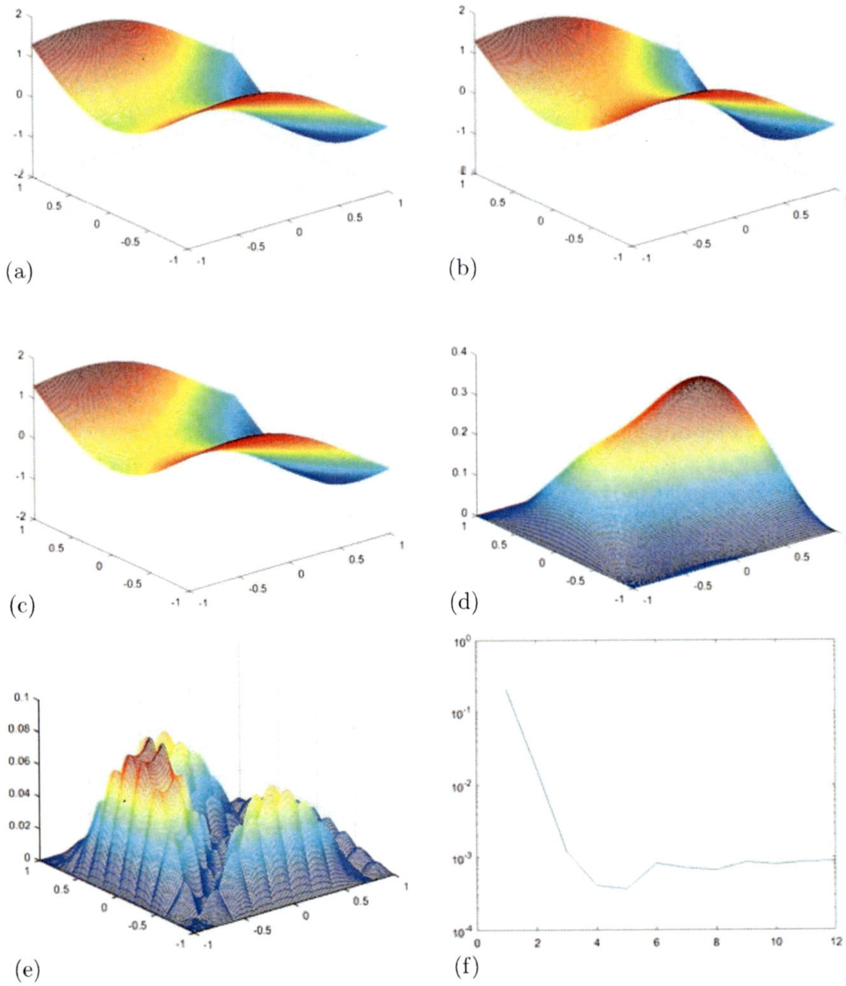


**Fig. 3** (Section 4.2) **a** The exact non-constant potential $q$, **b** the initial condition for $q$, **c** the inverted solution with full measurements on the boundary, the absolute error in **d** the initial potential $q$ and **e** the final potential $q$, and **f** the convergence history

To better investigate how the number of measurements improve the inversion results, we show in Table 1 some detailed convergence results. We run the same test case with different number of measurements varying from only 40 to the full measurements containing 640 measurements. As expected, the more information we obtained, the better the accuracy in the numerical solution. We obtain a better approximation of the potential $q$ with more boundary measurements of the DN map.

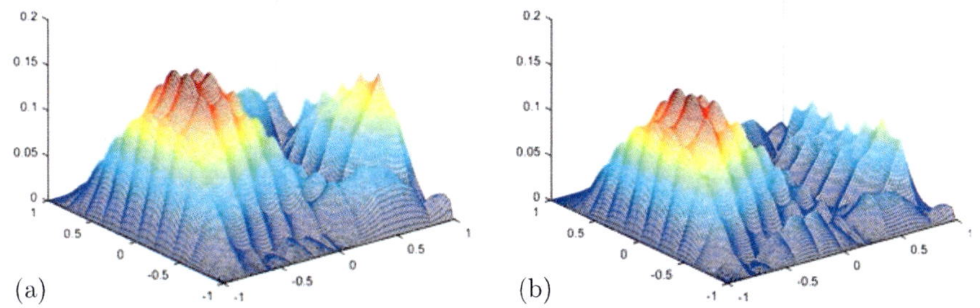


**Fig. 4** (Section 4.2) The absolute error in the inverted potential $q$ using **a** 40 and **b** 80 measurements where we assume boundary value of the potential $q$ on the finest mesh

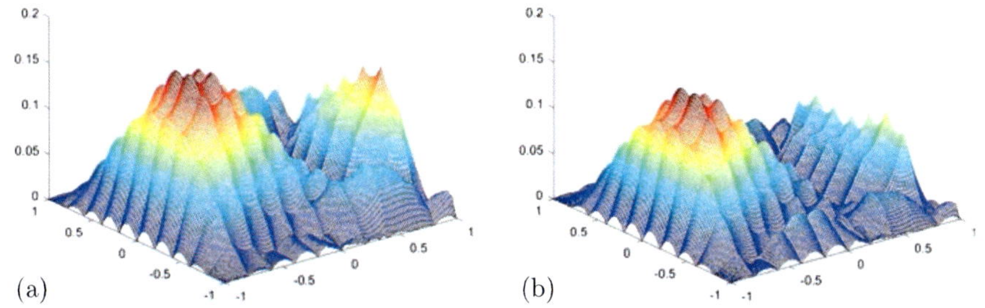


**Fig. 5** (Section 4.2) The absolute error in the inverted potential $q$ using **a** 40 and **b** 80 measurements where we assume boundary value of the potential $q$ on the coarsest mesh

Further, we observe that it actually takes more iterations for the case with finite number of measurements to converge to the steady state solution.

In the above tests, we have assumed full information of the boundary potential $q^*$. In practice, however, we might not be able to provide full measurements of the Schrödinger potential $q$ on the boundary. We show in Fig. 5 the inverted results of the potential $q$ given 40 and 80 measurements of the DN map respectively, given Dirichlet boundary condition of the potential $q$ only on the coarsest mesh with mesh size $11 \times 11$. Since the boundary value of $q$ is given by interpolation on finer mesh, we can see errors on the boundary, with only 40 points fixed. However, number of measurements of $q$ does not influence the error in the potential $q$ inside the domain, which means that our algorithm could recover the Schrödinger potential given only finite number of boundary measurements of $q$.

To test the stability of the algorithm derived by the adjoint state method, we add a multiplicative Gaussian noise to the DN map $\partial_{\mathbf{n}} u^*$ on the boundary $\partial\Omega$ and investigate the corresponding effect on the inverted potential. To see the influence of noise on the inverted results, we perform our algorithm on cases with different amount of Gaussian noise until the steady state and compare the $L_2$- error in the

**Table 1** (Section 4.2) The required number of iterations and the error in the solutions using different number of DN map measurements

| Number of measurements | Number of iterations | $L^2$-error in the inverted $q$ |
|---|---|---|
| 40 | 75 | 0.2106 |
| 80 | 72 | 0.1451 |
| full (640) | 11 | 0.0941 |

**Table 2** (Section 4.2) The required number of iterations and the error in the solutions when the DN map measurements contain noise

| Amount of gaussian noise | Number of iterations | $L^2$ error in $q$ | The mismatching energy |
|---|---|---|---|
| Clean measurements | 11 | 0.0941 | 8.20e-04 |
| 1% noise | 13 | 0.1019 | 0.1930 |
| 5% noise | 12 | 0.1107 | 0.2330 |
| 10% noise | 15 | 0.1212 | 0.3600 |

potential $q$. Table 2 shows these errors with 1–10% noise in the measurement. We find that as we increase the amount of Gaussian noise, the $L^2$-error in the solution $q$ increases as well as the mismatching energy.

Finally, to see the improvement in the computational efficiency of our cascadic initialization approach as described in Algorithm 2, we compare the number of the iterations required to reach the steady state solution obtained by Algorithm 1 and Algorithm 2 using the same stopping criteria given by $\left|E^k - E^{k-1}\right| \leq 10^{-4}$. Since Algorithm 2 requires to solve several more elliptic equations in order to update $u(\mathbf{x})$ corresponding to the Schrödinger potential $q(\mathbf{x})$ after interpolation in each level, we also look into the overall computation time. We find that Algorithm 1 requires 26 iterations taking 157.971 s of the CPU time, while Algorithm 2 takes only 11 iterations in total requiring only 2.472 s to reach the final solution. This shows that the cascadic initialization strategy as developed in Algorithm 2 can significantly improve the computational efficiency of the inversion method.

### *4.3 Discontinuous Models*

In all discussions above, we have assumed that the potential $q$ is smooth so that one can impose the Tikhonov regularization in the inversion process. In some applications, on the other hand, the unknown potential might actually violate such a strong regularity. In this example, we test the limit of the proposed approach by considering two examples where the exact potential is discontinuous.

#### 4.3.1 $q = \chi_{\mathbb{R}^+}$

We first consider the case where $q$ is piecewise continuous to see that whether our algorithm could recover the interface. In this example, we consider $q = \chi_{\mathbb{R}^+}$ in the domain $[-1, 1]^2$ with the Dirichlet boundary condition $u^*|_{\partial\Omega} = \sin x \sin y$. We stop the iterations when $|E^k - E^{k-1}| \leq 10^{-6}$. Figure 6 shows our inverted results given by full measurements and 80 measurements of the DN map on the boundary, respectively. Clearly we are not able to exactly recover the discontinuous potential but can roughly invert the macroscopic structure in the interior.

#### 4.3.2 A Circular Discontinuity

In the previous test, the potential measurements itself contain discontinuity. In this case, the discontinuity locates completely inside the domain instead of on the boundary. In this section, we assume the exact potential is given by

$$q(r) = \begin{cases} 1, & r \leq a \\ 0, & r > a. \end{cases}$$

We vary the radius $a$ in order to see the effect of distance between the discontinuity interface and the boundary. We consider the domain $[-1, 1]^2$ and impose the Dirichlet boundary condition $u^*|_{\partial\Omega} = \exp(x + y)$. Since the Schrödinger potential $q$ is not smooth in this case, we set the regularization parameter $\nu = 0.1$ and we stop the iterations when $|E^k - E^{k-1}| < 10^{-4}$. Figures 7 and 8 show the inverted results with radius 1 given full and finite measurements of the DN map on the boundary, respectively. We can see that the circular interface is roughly recovered and the inverted results preserve the symmetry of $q$.

Figure 8 shows the results when we set the radius to be 0.8, 0.6, and 0.4, respectively, while all other parameters in the model remain the same. Since the boundary value of $q$ is 0, our initial approximation of $q$ is constant 0, which is the solution to the Laplace equation with Dirichlet boundary condition $q^*|_{\partial\Omega} = 0$. We can see that as the distance between the discontinuity interface and the boundary increases, the magnitude of inverted results decrease although the shape and location of the interface can be very roughly recovered.

## 5 Conclusion

We have proposed a numerical approach for solving an inverse problem for the Schrödinger equation. The method is developed based on a least squares fitting functional where the gradient is computed using the adjoint state method. To further improve the computational efficiency of the gradient descent approach, we have

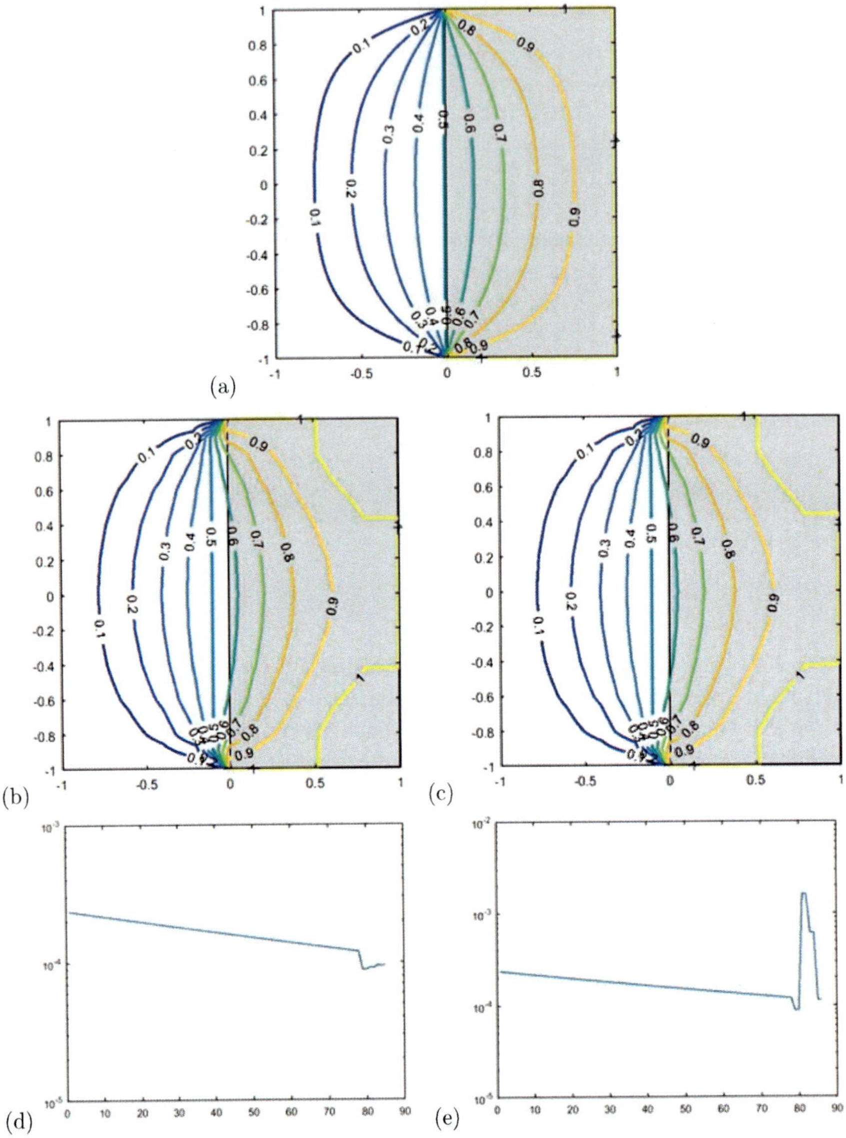


**Fig. 6** (Section 4.3.1) **a** The initial condition for the potential $q$. Our inverted potential using **b** full and **c** 80 measurements and their convergence history in (**d**) and (**e**), respectively

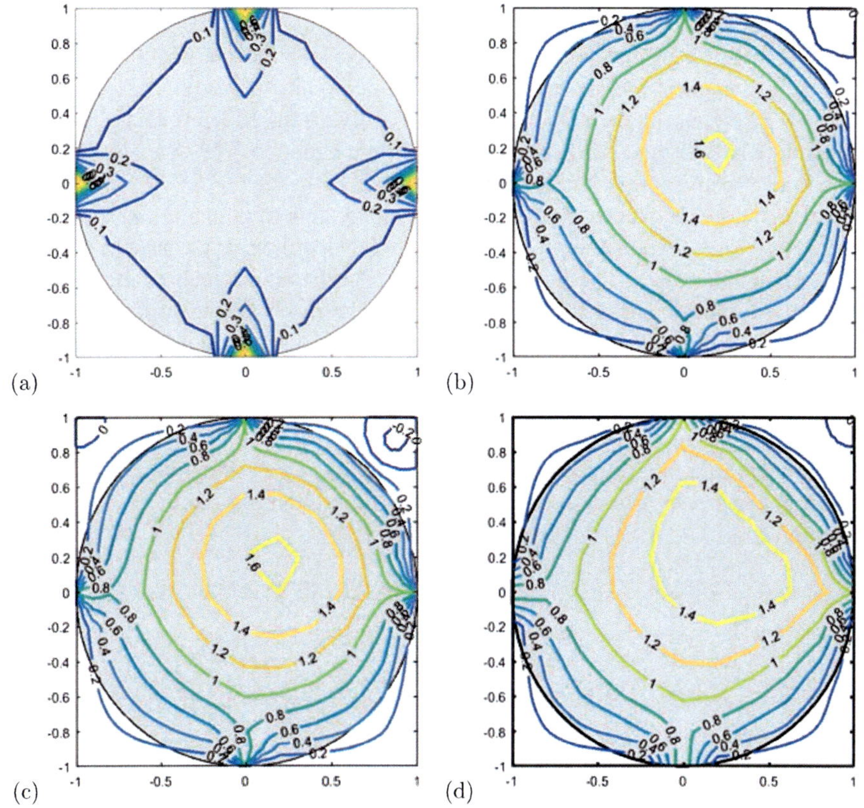


**Fig. 7** (Section 4.3.2) A circular discontinuity model with radius 1. The initial approximation **a**, the inverted results obtained by **b** full measurements, **c** 80 measurements, and **d** 20 measurements

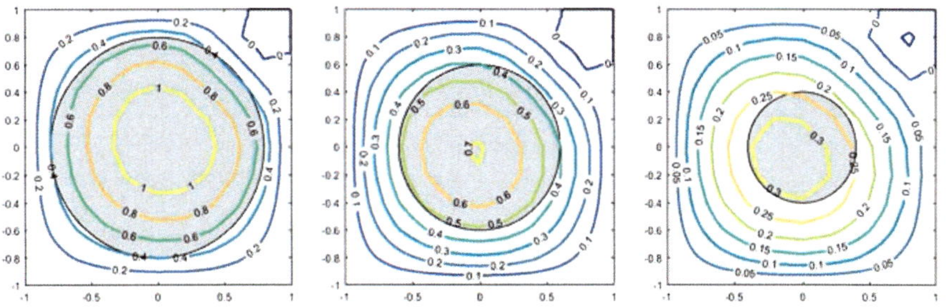

**Fig. 8** (Section 4.3.2) A circular discontinuity model with full measurements. The radius of the discontinuity is given by (left) 0.8, (middle) 0.6 and (right) 0.4

also developed an efficient cascadic initialization approach to reduce the number of iterations. Numerical results showed the efficiency, stability and fast convergence behavior of the energy using our method. The method developed in this paper can also be regarded as an alternative approach for solving the EIT problem with extra computational cost of solving one standard elliptic equation. We have also demonstrated the effectiveness of the approach in Sect. 4.2.

Indeed the method has been developed based on smooth potentials $q$ because of the regularity imposed in computing the gradient. Nevertheless, we have also tested on discontinuous cases. The inverted solutions clearly cannot recover the location of the discontinuity, but is still able to estimate the macroscopic structures in the potential. To better solve the inverse problem with a discontinuous potential, one might incorporate the level set method [31–33] in the adjoint state method as in [25, 26] for transmission traveltime tomography or in [19, 20, 27] for inverse gravimetry.

A possible future research direction is to apply this inverse solver for the Schrödinger equation to the EIT problem and to investigate how the efficiency of the current approach can help solving the inverse problem in the field of medical imaging.

**Acknowledgements** The work of Leung was supported in part by the Hong Kong RGC grant 16302819.

## References

1. G.S. Alberti, M. Santacesaria, Calderóns' inverse problem with a finite number of measurements. *Forum of Mathematics, Sigma*, vol. 7(e35) (2019)
2. S. Arridge, S. Moskow, C. Schotland, Inverse born series for the calderon problem. Inverse Probl. **28** (2012)
3. E. Beretta, M.V. de Hoop, L. Qiu, Lipschitz stability of an inverse boundary value problem for a Schrödinger type equation. SIAM J. Math. Anal. **45**(2), 679–699 (2013)
4. E. Beretta, S. Micheletti, S. Perotto, M. Santacesaria, Reconstruction of a piecewise constant conductivity on a polygonal partition via shape optimization in EIT. J. Comput. Phys. **353**, 264–280 (2018)
5. L. Borcea, Electrical impedance tomography. Inverse Probl. **18**, R99–R136 (2002)
6. R.M. Brown, G. Uhlmann, Uniqueness in the inverse conductivity problem for nonsmooth conductivities in two dimensions. Commun. Partial diff. Equ. **22**(5–6), 1009–1027 (1997)
7. M. Brühl, M. Hanke, M.S. Vogelius, A direct impedance tomography algorithm for locating small inhomogeneities. Numerische Mathematik **93**, 635–654 (2003)
8. A.P. Calderón, *On an Inverse Boundary Value Problem, Seminar on Numerical Analysis and its Applications to Continuum Physics* (Soc. Brasil. Mat, Rio de Janeiro, 1980), pp. 65–73
9. T.F. Chan, X.-C. Tai, Level set and total variation regularization for elliptic inverse problems with discontinuous coefficients. J. Comput. Phys. **193**, 40–66 (2003)
10. Z. Chen, J. Zou, An Augmented Lagrangian method for identifying discontinuous parameters in elliptic systems. SIAM J. Control Optim. **37**(3), 892–910 (1999)
11. M. Cheney, D. Isaacson, J.C. Newell, Electrical impedance tomography. SIAM Rev. **41**(1), 85–101 (1999)
12. M. Cheney, D. Isaacson, J.C. Newell, S. Simske, J. Goble, NOSER: an algorithm for solving the inverse conductivity problem. Int. J. Imaging Syst. Technol. **2**(2), 66–75 (1990)

13. E.T. Chung, T.F. Chan, X.-C. Tai, Electrical impedance tomography using level set representation and total variational regularization. J. Comput. Phys. **205**(1), 357–372 (2005)
14. P.M. Edic, D. Isaacson, G.J. Saulnier, H. Jain, J.C. Newell, An iterative Newton-Raphson method to solve the inverse admittivity problem. IEEE Trans. Biomed. Eng. **45**(7), 899–908 (1998)
15. A. Friedman, V. Isakov, On the uniqueness in the inverse conductivity problem with one measurement. Indiana Univ. Math. J. **38**(3), 563–579 (1989)
16. R. Glowinski, S. Leung, J. Qian, A simple explicit operator-splitting method for effective Hamiltonians. SIAM J. Sc. Comput. **40**(1), A484–A503 (2018)
17. S.J. Hamilton, EIT imaging of admittivities with a D-bar method and spatial prior: experimental results for absolute and difference imaging. Physiol. Meas. **38**(6), 1176–1192 (2017)
18. A. Hauptmann, M. Santacesaria, S. Siltanen, Direct inversion from partial-boundary data in electrical impedance tomography. Inverse Probl. **33**(025009) (2017)
19. V. Isakov, S. Leung, J. Qian, A fast local level set method for inverse gravimetry. Commun. Comput. Phys. **10**, 1044–1070 (2011)
20. V. Isakov, S. Leung, J. Qian, A three-dimensional inverse gravimetry problem for INCE with snow caps. Inverse Probl. Imaging **7**(2), 523–544 (2013)
21. J.P. Kaipio, V. Kolehmainen, E. Somersalo, M. Vauhkonen, Statistical inversion and Monte Carlo sampling methods in electrical impedance tomography. Inverse Probl. **16**(5), 1487–1522 (2000)
22. K. Knudsen, M. Lassas, J.L. Mueller, S. Siltanen, D-bar method for electrical impedance tomography with discontinuous conductivities. SIAM J. Appl. Math. **67**(3), 893–913 (2007)
23. A. Lechleiter, A. Rieder, Newton regularizations for impedance tomography: a numerical study. Inverse Probl. **22**, 1967–1987 (2006)
24. S. Leung, J. Qian, An adjoint state method for 3D transmission traveltime tomography using first arrival. Commun. Math. Sci. **4**, 249–266 (2006)
25. W.B. Li, S. Leung, A fast local level set adjoint state method for first arrival transmission traveltime tomography with discontinuous slowness. Geophs. J. Int. **195**(1), 582–596 (2013)
26. W.B. Li, S. Leung, J. Qian, A level-set adjoint-state method for crosswell transmission-reflection traveltime tomography. Geophs. J. Int. **199**(1), 348–367 (2014)
27. W. Lu, S. Leung, J. Qian, An improved fast local level set method for three-dimensioanl inverse gravimetry. Inverse Probl. Imaging **9**(2), 479–509 (2015)
28. N. Manache, Exponential instability in an inverse problem for the Schrödinger equation. Inverse Probl. **17**, 1435–1444 (2001)
29. J.L. Muller, S. Siltanen, *Linear and Nonlinear Inverse Problems With Practical Applications* (SIAM, 2012)
30. A.I. Nachman, Global uniqueness for a two-dimensional inverse boundary value problem. Ann. Math. **143**(1), 71–96 (1995)
31. S.J. Osher, R.P. Fedkiw, *Level Set Methods and Dynamic Implicit Surfaces* (Springer, New York, 2003)
32. S.J. Osher, J.A. Sethian, Fronts propagating with curvature dependent speed: algorithms based on Hamilton-Jacobi formulations. J. Comput. Phys. **79**, 12–49 (1988)
33. J.A. Sethian. *Level Set Methods* (2nd edn., Cambridge University Press, 1999)
34. J. Sylvester, G. Uhlmann, A global uniqueness theorem for an inverse boundary value problem. Ann. Math. **125**(1), 153–169 (1987)
35. G. Uhlmann, Commentary on Calderón's paper (29), on an inverse boundary value problem. *Selected papers of Alberto P. Calderón*, pp. 623–636 (2008)
36. G. Uhlmann, Electrical impedance tomography and Calderón's problem. Inverse Probl. **25**(12) (2009)
37. G. Uhlmann, Inverse problems: seeing the unseen. Bull. Math. Sci. **4**, 209–279 (2014)

# Multi-modality Image Registration Models and Efficient Algorithms

**Daoping Zhang, Anis Theljani, and Ke Chen**

*Dedicated to Professor R. H. Chan on the Occasion of His 60th Birthday.*

**Abstract** In this Chapter we discuss multi-modality image registration models and efficient algorithms. We propose a simple method to enhance a variational model to generate a diffeomorphic transformation. The idea is illustrated by using a particular model based on reformulated normalized gradients of the images as the fidelity term and higher-order derivatives as the regularizer. By adding a control term motivated by quasi-conformal maps and Beltrami coefficients, the model has the ability to guarantee a diffeomorphic transformation. Without this feature, the model may lead to visually pleasing but invalid results. To solve the model numerically, we present both a Gauss-Newton method and an augmented Lagrangian method to solve the resulting discrete optimization problem. A multilevel technique is employed to speed up the initialization and reduce the possibility of getting local minima of the underlying functional. Finally numerical experiments demonstrate that this new model can deliver good performances for multi-modal image registration and simultaneously generate an accurate diffeomorphic transformation.

**Keywords** Multi-modal image registration · Variational model · Diffeomorphic transformation

**AMS subject classifications** 65K10 · 68U10

D. Zhang · A. Theljani · K. Chen (✉)
EPSRC Liverpool Centre for Mathematics in Healthcare, Centre for Mathematical Imaging Techniques and Department of Mathematical Sciences, The University of Liverpool, Peach Street, Liverpool L69 7ZL, UK
e-mail: k.chen@liv.ac.uk
URL: https://www.liv.ac.uk/cmit



X.-C. Tai et al. (eds.), *Mathematical Methods in Image Processing and Inverse Problems*, Springer Proceedings in Mathematics & Statistics 360,
https://doi.org/10.1007/978-981-16-2701-9_3

## 1 Introduction

Working on a pair of images of the same object taken at different times or acquired using different devices, image registration aims to either find differences between them or fuse complementary information to each other which is otherwise not possible with a single modality. In either case, the key is to find a reasonable spatial geometric transformation between these two images. Though the task is required in diverse fields such as astronomy, optics, biology, chemistry and remote sensing and particularly in medical imaging, and much work have been done, getting a robust model for the task is still a challenge. For an overview of image registration methodologies and approaches, especially for registering images acquired by the same modality (e.g. CT-CT), we refer to [17, 18, 33, 35, 40]. For a more recent survey, see [8]. This Chapter is mainly concerned with registering two images from different modalities (e.g. CT-MRI or digital-Infrared) and focuses on one important question of how to impose a constraint so that the underlying transformation is diffeomorphic.

The image registration problem can be described as follows: given a fixed image $R$ (the reference) and a moving image $T$ (the template), both represented by scalar function mappings over $\Omega \subset \mathbb{R}^d \longrightarrow \mathbb{R}$, find a suitable geometric transformation $\varphi(\boldsymbol{x}) = \boldsymbol{x} + \boldsymbol{u}(\boldsymbol{x}),\ \boldsymbol{u} : \mathbb{R}^d \longrightarrow \mathbb{R}^d$ such that

$$G_1(T[\varphi]) = G_1(T(\boldsymbol{x} + \boldsymbol{u}(\boldsymbol{x})) \approx G_2(R), \tag{1}$$

where $G_1, G_2$ must be chosen suitably in multi-modality scenario, because only features or patterns in $T, R$ visually resemble each other, not their given intensities. In contrast, in mono-modality registration where intensities as well as features in $T, R$ resemble each other, we have $G_i(\cdot) = I_d,\ (i = 1, 2)$ or $T \approx R$ pixel wise. In the special case of parametric models, the solution $\boldsymbol{u}$ (or $\varphi$) is assumed to belong to some linear spanned space with known Ansatz functions, depending on few parameters (e.g. affine with 6 parameters in 2D or 12 parameters in 3D). However, not all problems can be solved by parametric models.

Here, we focus on variational models for deformable non-parametric image registration where the unknown $\boldsymbol{u}$ sought in a properly chosen functional space is not assumed to have any parametric forms. The reconstruction problem based on model (1) is an ill-posed inverse problem and thus regularization techniques are needed to overcome ill-posedness [7, 11, 13, 14, 21, 30, 31, 47]. Generally speaking, a regularization technique turns the ill-posed problem (1) into a well-posed optimization model

$$\min_{\mathbf{u} \in \mathcal{H}} \left\{ \mathcal{J}(\mathbf{u}) = S(\boldsymbol{u}) + \frac{\lambda}{2} D(T(\boldsymbol{x} + \boldsymbol{u}), R) \right\} \tag{2}$$

where the displacement $\boldsymbol{u}$ is a minimizer of the above joint energy functional and $\lambda$ is a positive weight which controls the trade-off between them.

In (2), the **first term** $S(\boldsymbol{u})$ is a regularization term which controls the smoothness of $\boldsymbol{u}$ and reflects our expectations in penalizing unlikely transformations. Various regularizers have been proposed, such as first-order derivatives-based on total variation [10, 23], diffusion [15] and elastic regularizer registration models, higher-order derivatives-based on linear curvature [16], mean curvature [12], Gaussian curvature [24], and fractional order derivatives based models [50]; refer also to [11, 31, 44, 51, 52].

The **second term** $D(T(\boldsymbol{x}+\boldsymbol{u}), R)$ is a fidelity measure, which quantifies distance or similarity between the transformed template image $T(\boldsymbol{x}+\boldsymbol{u})$ and the reference $R$. For mono-modal registration, a widely-used data fidelity term $D(T(\boldsymbol{x}+\boldsymbol{u}), R)$ is the sum of squared differences $D = \|T(\boldsymbol{x}+\boldsymbol{u}) - R\|_2^2 \equiv \text{SSD}(T(\boldsymbol{x}+\boldsymbol{u}), R)$ to measure the difference between the reference image $R$ and the deformed template image $T(\boldsymbol{x}+\boldsymbol{u})$. However for multi-modality registration, the choice of $D(T(\boldsymbol{x}+\boldsymbol{u}), R)$ is more challenging. The main issue is how to design the right (or rather better) similarity measures that can support the difference (in features, colours, gradients, illumination etc.) between images from different modalities (e.g. SSD no longer makes sense). Various measures have been proposed and tested in the literature. Designing a measure which is based on the geometric information such as the gradients of the images is a good choice. See for instance the normalized gradient field (**NGF**) [22, 26, 39], edges sketching registration [1], normalized gradient fitting (**GT**) [22, 43] and Mutual Information [29, 37, 46]. Recently [9] proposed a cross-correlation similarity measure based on reproducing kernel Hilbert spaces and found advantages over Mutual Information.

Many models in the literature, of type (2), do not usually contain constraints to ensure that $\varphi(\boldsymbol{x})$ is a diffeomorphic map for the mono-modal registration. And even fewer theoretical or experimental studies deal with diffeomorphic maps for the multi-modal registration. But non-diffeomorphic maps cause phenomena such as folding or tearing which are usually seen as non-natural transformations between the two images, unless $\lambda$ is small (implying a poor registration fidelity error). Over the last decade, more and more researchers have focused on diffeomorphic image registration where folding measured by the local invertibility quantity $\det(J_\varphi)$ is reduced or avoided where $\det(J_\varphi)$ is the Jacobian determinant of $\varphi$. Under desired assumptions, obtaining a one-to-one mapping is a natural choice, see [7, 14, 19, 20].

After surveying a few models of type (2) for multi-modal images, this Chapter shows how to incorporate a suitable constraint into a model so that it can deliver a diffeomorphic map. We illustrate our idea by a specific model: minimizing a new functional based on using reformulated normalized gradients of the images as the fidelity term [43], higher-order derivatives and a new Beltrami coefficient based term [28, 48]. An effective, iterative scheme is also presented and numerical experimental results show that the new registration model has a good performance.

## 2 Review of Related Models

For a variational image registration model (2), while there exist many choices for a regularizer $S(\boldsymbol{u})$ such as the diffusion operator or the Laplacian [8], below, we briefly review a few of such choices of $D(T(\boldsymbol{x}+\boldsymbol{u}), R)$ for registering a pair of multi-modal images $T, R$.

**Normalized Gradient Field (NGF) and its variants**. The basic idea of **NGF** [22, 26, 39] is the use of a derived information from the image intensity, i.e., the gradient. Similarity measures depending on the gradients or geometry of the images, which naturally encode information about the shape, can be better. The aim is to align the gradients $\nabla T(\boldsymbol{x}+\boldsymbol{u})$ and $\nabla R$ by minimizing the cosines distance between them. More precisely, on each point $\boldsymbol{x} \in \Omega$, try to find a displacement $\boldsymbol{u}(x)$ such that $\cos\Theta = 1$ where $\Theta$ is the angle between $\nabla T(\boldsymbol{x}+\boldsymbol{u})$ and $\nabla R$, which leads to minimizing the similarity term:

$$D^{NGF}(T(\boldsymbol{x}+\boldsymbol{u}), R) = \int_\Omega (1-(\cos\Theta)^2)\,\mathrm{d}\boldsymbol{x} = \int_\Omega (1-(\nabla_n T(\boldsymbol{x}+\boldsymbol{u})\cdot\nabla_n R)^2)\,\mathrm{d}\boldsymbol{x}, \tag{3}$$

where $\nabla_n T(\boldsymbol{x}+\boldsymbol{u}) = \nabla T(\boldsymbol{x}+\boldsymbol{u})/|\nabla T(\boldsymbol{x}+\boldsymbol{u})|$ and $\nabla_n R = \nabla R/|\nabla R|$ are normalized unit vectors. An alternative form of the **NGF** that avoids using terms $\nabla_n T(\boldsymbol{x}+\boldsymbol{u})$ and $\nabla_n R$ which are degenerated in homogeneous regions, reformulate **NGF** as

$$D^{NGF}(T(\boldsymbol{x}+\boldsymbol{u}), R) = \int_\Omega (|\nabla T(\boldsymbol{x}+\boldsymbol{u})|^2|\nabla R|^2 - (\nabla T(\boldsymbol{x}+\boldsymbol{u})\cdot\nabla R)^2)\,\mathrm{d}\boldsymbol{x}, \tag{4}$$

**Mutual Information (MI)**. It was firstly proposed in [46] and has been studied in various literatures (see [29, 37]), showcasing its great capability as well as limitations. The basic idea is to compare the histograms of the images by exploiting the following quantity

$$D^{MI}(T(\boldsymbol{x}+\boldsymbol{u}), R) = -\int_{\mathbb{R}^2} p_{T,R}(t,r)\log\frac{p_{T,R}(t,r)}{p_T(t)p_R(r)}\,\mathrm{d}t\mathrm{d}r, \tag{5}$$

where $p_R, p_T$ are probability distributions of the gray values in $R$ and $T$, while $p_{T,R}$ is the joint probability of the gray values which can be derived from the joint histogram. The main drawback of **MI** is its sensibility to image quantization and the difficulty in estimating the joint probability density function (PDF). In addition, the measure also fails when two features with different intensities in one image have similar intensities in the other one [27].

**Maximum Correlation Coefficient (MCC)**. It is an extension of well-known Normalized cross correlation (**CC**) measure , which is only efficient for mono-modal images [6, 33], to a measure that is able to handle multi-modal images [9]. The similarity measure is defined by

$$D^{MCC}(T(x+u), R) = (1 - \mathbf{MCC}(T, R))^p := (1 - \max_{f,g} \mathbf{CC}(M, N))^p, \; 0 < p < 1,$$

where $M(x) = f(T(x+u), N(x) = g(R(x)), f$ and $g$ are two measurable functions. This **MCC** formulation does not require estimation of the continuous joint PDF and offers a powerful alternative to the models based on maximizing **MI**. However. the computation of the maximum over all functions $f$ and $g$ is a big challenge. The recommended approach in [9] is to approximate it based on the theory of reproducing kernel Hilbert space (**RKHS**) [2, 5].

# 3 The New Model

We aim to design a variational model building on the energy of the form (2)

$$\min_{\mathbf{u}\in\mathcal{H}} \left\{ \mathcal{J}(\mathbf{u}) = S(u) + D(T(x+u), R) + \gamma C(u) \right\} \tag{6}$$

which is comprised of three building blocks: a data fidelity term with similarity measure $D$, a regularization term $S$ and a control term $C$. The emphasis of this Chapter is how to choose $C$. To do this for a concrete model, we now specify our choice of all three terms.

## 3.1 Data Fitting

We consider a similarity measure based on the gradient information [43]. This measure is motivated by the standard **NGF** [22, 32] and it primarily explores the potential of normalized gradients beyond its standard form. We shall consider normalized gradients fitting combined with a measure based on the triangular similarity inequality. More precisely, we consider the following fitting term

$$D(T(x+u), R) = D^{GF}(u) + \alpha D^{TM}(u) \tag{7}$$

where GF stands for 'gradient filed difference' and TM for 'Triangular Measure' with

$$D^{GF}(u) = \int_{\Omega} |\nabla_n T(x+u) - \nabla_n R|^2 \mathrm{d}x,$$

$$D^{TM}(u) = \int_{\Omega} (|\nabla T(x+u)| + |\nabla R| - |\nabla T(x+u) + \nabla R|)^2 \mathrm{d}x.$$

## 3.2 *Regularization*

A regularizer controls the smoothness. Our primary choice for smoothness control is the diffusion model [15] which uses first-order derivatives and promotes smoothness. As affine linear transformations are not included in the kernel of the $H^1$-regularizer, we desire a regularizer which can penalize such transformation. As such, we add the regularizer based on second-order derivatives (LLT) to the model which allows to remove the need of any pre-registration step of affine transformations. The second-order derivatives allows also getting smooth transformations [52]. Our adopted regularizer is given by

$$S(\boldsymbol{u}) = \frac{\beta_1}{2} S_1(\boldsymbol{u}) + \frac{\beta_2}{2} S_2(\boldsymbol{u}) \tag{8}$$

where

$$S_1(\boldsymbol{u}) = \int_{\Omega} |\nabla \boldsymbol{u}|^2 \mathrm{d}\boldsymbol{x}, \qquad S_2(\boldsymbol{u}) = \int_{\Omega} |\nabla^2 \boldsymbol{u}|^2 \mathrm{d}\boldsymbol{x}.$$

## 3.3 *Invertibility*

A diffeomorphic map ensures local invertibility of the map and this is achievable by a control term $C$ that imposes the constraint $\det(J_\varphi) > 0$ at any $\boldsymbol{x} \in \Omega$. This latter idea is much used in the literature with somewhat limited success because either strong assumptions on $T$, $R$ or compromised fidelity error are required; see tests and remarks from [48]. Here, instead of controlling $\det(J_\varphi)$ directly, we control the Beltrami coefficient [48] in getting a diffeomorphic map and propose the use of

$$C(\boldsymbol{u}) = \int_{\Omega} \phi(|\mu(\boldsymbol{u})|^2) \mathrm{d}\boldsymbol{x}, \tag{9}$$

where $\phi(v) = \frac{v^2}{(v-1)^2}$ and $|\mu(\boldsymbol{u})|^2 = \frac{(\partial_{x_1} u_1 - \partial_{x_2} u_2)^2 + (\partial_{x_2} u_1 + \partial_{x_1} u_2)^2}{(\partial_{x_1} u_1 + \partial_{x_2} u_2 + 2)^2 + (\partial_{x_2} u_1 - \partial_{x_1} u_2)^2}$.

One notes that our choice of the first two terms $S$, $D$ for (6) is quite common while the third term [48] is relatively new to readers. This is the key idea of this Chapter: an old, non-diffeomorphic, variational model of form (2) can be converted to a diffeomorphic model by adding a control term such as $C$ from (9). This can be done in 2D and also in 3D following our recent work. It should be remarked that model (6) is non-convex so its solutions are not unique (as true for all registration models). However we can show that the model admits at least one solution in the space $W^{2,2}(\Omega)$, following the idea of [49].

# 4 The Solution Algorithm

Here, we choose first-discretize-then-optimize method, namely directly discretize the variational model to get a discrete optimization problem and then use optimization methods to solve this resulting optimization problem. In this section we focus on a Gauss-Newton (G-N) method and in the next section we briefly introduce another alternating iteration method just before numerical results are shown.

## 4.1 *Discretization*

In the implementation, we employ the nodal grid and define a spatial partition

$$\Omega_h^n = \{\boldsymbol{x}^{i,j} \in \Omega | \boldsymbol{x}^{i,j} = (x_1^i, x_2^j) = (ih, jh), 0 \le i \le n, 0 \le j \le n\},$$

where $h = \frac{1}{n}$ and the discrete domain consists of $n^2$ cells of size $h \times h$. We discretize the displacement field $\boldsymbol{u}$ on the nodal grid, namely $\boldsymbol{u}^{i,j} = (u_1^{i,j}, u_2^{i,j}) = (u_1(x_1^i, x_2^j), u_2(x_1^i, x_2^j))$. By lexicographical ordering, we reshape four matrices to two long vectors of dimension $\mathbb{R}^{2(n+1)^2 \times 1}$

$$X = (x_1^0, x_1^1, ..., x_1^n, \dots, x_1^0, x_1^1, ...x_1^n, x_2^0, x_2^0, ..., x_2^0, \dots, x_2^n, x_2^n, ...x_2^n)^T,$$
$$U = (u_1^{0,0}, ..., u_1^{n,0}, \dots, u_1^{0,n}, ..., u_1^{n,n}, u_2^{0,0}, ..., u_2^{n,0}, \dots, u_2^{0,n}, ..., u_2^{n,n})^T.$$

### 4.1.1 Discretization of Fitting Term

Firstly, set $\mathbf{R} = \mathbf{R}(PX) \in \mathbb{R}^{n^2 \times 1}$ as the discretized reference image and $\mathbf{T}(PX + PU) \in \mathbb{R}^{n^2 \times 1}$ as the discretized deformed template image, where $P \in \mathbb{R}^{2n^2 \times 2(n+1)^2}$ is an average matrix from the nodal grid to the cell-centered grid. In order to discretize $\nabla T$ and $\nabla R$, we introduce two discrete operators: $D_1 = I_n \otimes \partial_h^1$ and $D_2 = \partial_h^1 \otimes I_n$, where

$$\partial_h^1 = \frac{1}{2h} \begin{bmatrix} -1 & 1 & & & \\ -1 & 0 & 1 & & \\ & \dots & \dots & \dots & \\ & & -1 & 0 & 1 \\ & & & -1 & 1 \end{bmatrix} \in \mathbb{R}^{n \times n}.$$

Hence, the discretized $\nabla T$ and $\nabla R$ are $[D_1\mathbf{T}, D_2\mathbf{T}]$ and $[D_1\mathbf{R}, D_2\mathbf{R}]$ respectively. Set $\mathrm{LT} = (\sum_{i=1}^2 D_i\mathbf{T} \odot D_i\mathbf{T} + \epsilon)^{\cdot 1/2}$, $\mathrm{LR} = (\sum_{i=1}^2 D_i\mathbf{R} \odot D_i\mathbf{R} + \epsilon)^{\cdot 1/2}$ and $\mathrm{LTR} = (\sum_{i=1}^2 D_i(\mathbf{T} + \mathbf{R}) \odot D_i(\mathbf{T} + \mathbf{R}) + \epsilon)^{\cdot 1/2}$, where $\odot$ indicates component-wise product and $(\cdot)^{\cdot 1/2}$ indicates the component-wise square root.

Then for $D^{GF}(\boldsymbol{u})$ and $D^{TM}(\boldsymbol{u})$, we have the following discretizations:

$$D^{GF}(\boldsymbol{u}) \approx h^2 p_1^T p_1, \quad D^{TSM}(\boldsymbol{u}) \approx h^2 p_2^T p_2, \tag{10}$$

where (using ./ to indicate the component-wise division)

$$p_1 = [D_1\mathbf{T}./\mathrm{LT} - D_1\mathbf{R}./\mathrm{LR};\ D_2\mathbf{T}./\mathrm{LT} - D_2\mathbf{R}./\mathrm{LR}]$$
$$p_2 = \mathrm{LT} + \mathrm{LR} - \mathrm{LTR}.$$

### 4.1.2 Discretization of Regularization Term

The first-order regularization term can be discretized into the following form:

$$S_1(\boldsymbol{u}) \approx h^2 \sum_{i=0}^{n-1}\sum_{j=0}^{n-1}\sum_{l=1}^{2} \left(\frac{u_l^{i+1,j} - u_l^{i,j}}{h}\right)^2 + \left(\frac{u_l^{i,j+1} - u_l^{i,j}}{h}\right)^2 \tag{11}$$

by using the forward difference and mid-point rule.

Define $B_1 = I_{n+1} \otimes \partial_h^2 \in \mathbb{R}^{(n+1)^2\times(n+1)^2}$, $C_1 = \partial_h^2 \otimes I_{n+1} \in \mathbb{R}^{(n+1)^2\times(n+1)^2}$,

$$\partial_h^2 = \frac{1}{h}\begin{bmatrix} -1 & 1 & & \\ \cdots & \cdots & \cdots & \\ & -1 & 1 & \\ & & 0 & \end{bmatrix} \in \mathbb{R}^{(n+1)\times(n+1)}, \quad A_1 = \begin{bmatrix} B_1 & 0 \\ C_1 & 0 \\ 0 & B_1 \\ 0 & C_1 \end{bmatrix} \in \mathbb{R}^{4(n+1)^2\times 2(n+1)^2},$$

where $\otimes$ denotes the Kronecker product. Then (11) can be rewritten into the following form (noting $U \in \mathbb{R}^{2(n+1)^2\times 1}$)

$$S_1(\boldsymbol{u}) \approx h^2 U^T A_1^T A_1 U. \tag{12}$$

The second-order regularization term can be discretized into the following:

$$S_2(\boldsymbol{u}) \approx h^2 \sum_{i=0}^{n-1}\sum_{j=0}^{n-1}\sum_{l=1}^{2} \left(\frac{u_l^{i+1,j} - 2u_l^{i,j} + u_l^{i-1,j}}{h^2}\right)^2 + \left(\frac{u_l^{i,j+1} - 2u_l^{i,j} + u_l^{i,j-1}}{h^2}\right)^2$$
$$+ 2h^2 \sum_{i=0}^{n-1}\sum_{j=0}^{n-1}\sum_{l=1}^{2} \left(\frac{u_l^{i,j} - u_l^{i+1,j} - u_l^{i,j+1} + u_l^{i+1,j+1}}{h^2}\right)^2 \tag{13}$$

by using the central difference, mid-point rule and Neumann boundary conditions ($l = 1, 2$): $u_l^{i,0} = u_l^{i,-1}$, $u_l^{i,n} = u_l^{i,n+1}$, $u_l^{0,j} = u_l^{-1,j}$, $u_l^{n,j} = u_l^{n+1,j}$.

Further define $B_{21} = I_2 \otimes (I_{n+1} \otimes \partial_h^3)$, $B_{22} = I_2 \otimes (\partial_h^3 \otimes I_{n+1})$, $C_2 = I_2 \otimes (E \otimes E)$, $\tau_1 = (n+1) \times (n+1)$, $\tau_2 = n \times (n+1)$, where

$$\partial_h^3 = \frac{1}{h^2}\begin{bmatrix} -1 & 1 & & & \\ 1 & -2 & 1 & & \\ & \cdots & \cdots & \cdots & \\ & & 1 & -2 & 1 \\ & & & 1 & -1 \end{bmatrix} \in \mathbb{R}^{\tau_1}, \quad E = \frac{1}{h}\begin{bmatrix} -1 & 1 & & & \\ & -1 & 1 & & \\ & \cdots & \cdots & \cdots & \\ & & & -1 & 1 \\ & & & & -1 \; 1 \end{bmatrix} \in \mathbb{R}^{\tau_2}.$$

Then (13) can be rewritten into the following form

$$S_2(\boldsymbol{u}) \approx h^2 U^T A_2 U, \quad A_2 = B_{21}^T B_{21} + B_{22}^T B_{22} + 2C_2^T C_2. \tag{14}$$

### 4.1.3 Discretization of Control Term

Note that $\phi(|\mu(\boldsymbol{u})|^2)$ involves only first order derivatives and all $\boldsymbol{u}^{i,j}$ are available at vertex pixels. Thus it is convenient first to obtain approximations at all cell centers (e.g. at $V_5$ in Fig. 1) and second to use local linear elements to facilitate first order derivatives. We shall divide each cell (Fig. 1) into 4 triangles. In each triangle, we construct two linear interpolation functions to approximate the $u_1$ and $u_2$. Consequently, all partial derivatives are locally constants or $\phi(|\mu(\boldsymbol{u})|^2)$ is constant in each triangle.

Set $\mathbf{L}^{i,j,k}(\boldsymbol{x}) = (L_1^{i,j,k}(\boldsymbol{x}), L_2^{i,j,k}(\boldsymbol{x})) = (a_1^{i,j,k}x_1 + a_2^{i,j,k}x_2 + a_3^{i,j,k}, a_4^{i,j,k}x_1 + a_5^{i,j,k}x_2 + a_6^{i,j,k})$, which is the linear interpolation for $\boldsymbol{u}$ in the $\Omega_{i,j,k}$. Note that $\partial_{x_1}L_1^{i,j,k} = a_1^{i,j,k}$, $\partial_{x_2}L_1^{i,j,k} = a_2^{i,j,k}$, $\partial_{x_1}L_2^{i,j,k} = a_4^{i,j,k}$ and $\partial_{x_2}L_2^{i,j,k} = a_5^{i,j,k}$. Then according to the partition in Fig. 1, we have

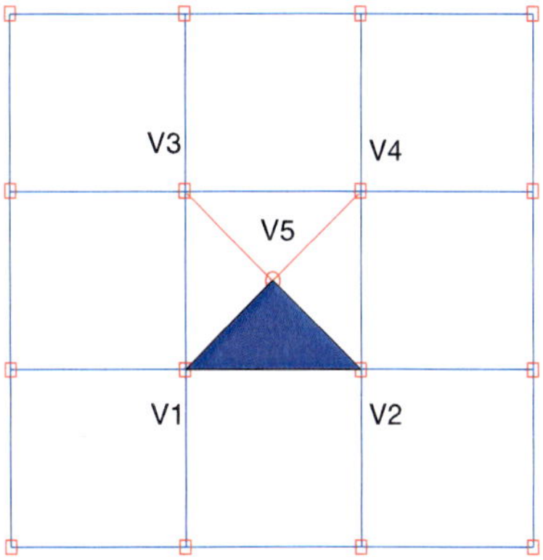


**Fig. 1** Partition of a cell, nodal point □ and center point ∘. $\triangle V_1 V_2 V_5$ is $\Omega_{i,j,k}$

$$
\begin{aligned}
C(\boldsymbol{u}) &= \int_{\Omega} \phi(|\mu(\boldsymbol{u})|^2)\mathrm{d}\boldsymbol{x} \\
&\approx \frac{h^2}{4}\sum_{i=1}^{n}\sum_{j=1}^{n}\sum_{k=1}^{4}\phi\Big(\frac{(a_1^{i,j,k}-a_5^{i,j,k})^2+(a_2^{i,j,k}+a_4^{i,j,k})^2}{(a_1^{i,j,k}+a_5^{i,j,k}+2)^2+(a_2^{i,j,k}-a_4^{i,j,k})^2}\Big).
\end{aligned}
\tag{15}
$$

To simplify (15), define 3 vectors $\mathbf{r}(U), \mathbf{r}^1(U), \mathbf{r}^2(U) \in \mathbb{R}^{4n^2}$ by $\mathbf{r}(U)_\ell = \mathbf{r}^1(U)_\ell \mathbf{r}^2(U)_\ell$, $\mathbf{r}^1(U)_\ell = (a_1^{i,j,k} - a_5^{i,j,k})^2 + (a_2^{i,j,k} + a_4^{i,j,k})^2$, $\mathbf{r}^2(U)_\ell = 1/[(a_1^{i,j,k} + a_5^{i,j,k} + 2)^2 + (a_2^{i,j,k} - a_4^{i,j,k})^2]$ where $\ell = (k-1)n^2 + (j-1)n + i \in [1, 4n^2]$.

Hence, (15) becomes

$$
C(\boldsymbol{u}) \approx \frac{h^2}{4}\phi(\mathbf{r}(U))e^T \tag{16}
$$

where $\phi(\mathbf{r}(U)) = (\phi(\mathbf{r}(U)_1), \ldots, \phi(\mathbf{r}(U)_{4n^2}))$ denotes the pixel-wise discretization of $u_1, u_2$ at all cell centers, and $e = (1, \ldots, 1) \in \mathbb{R}^{4n^2}$.

Finally, combining the above three parts (10), (12), (14) and (16), we get the discretization formulation for model (6):

$$
\min_U J(U) := h^2 p_1^T p_1 + \alpha h^2 p_2^T p_2 + \frac{\beta_1 h^2}{2} U^T A_1^T A_1 U + \frac{\beta_2 h^2}{2} U^T A_2 U + \frac{\gamma h^2}{4}\phi(\mathbf{r}(U))e^T. \tag{17}
$$

**Remark 1** According to the definition of $\phi$ and $\mathbf{r}(U)_\ell \geq 0$, each component of $\phi(\mathbf{r}(U))$ is non-negative and differentiable.

## 4.2 *Optimization Method for the Discretized Problem* (17)

In the numerical implementation, we choose a line search method to solve the resulting unconstrained optimization problem (17). Here, the basic iterative scheme is

$$
U^{i+1} = U^i + \theta \delta U^i, \tag{18}
$$

where $\delta U^i$ is the search direction and $\theta$ is the step length. In order to guarantee a descent search direction, we employ a Gauss-Newton method as the standard Newton method does not generate a descent direction because our exact Hessian is non-definite.

### 4.2.1 Gradient and Approximated Hessian of (17)

Firstly, we consider computing the gradient and approximated Hessian of the discretized fitting term $h^2 p_1^T p_1 + \alpha h^2 p_2^T p_2$. Its gradient and approximated Hessian are respectively:

$$\begin{cases} d_1 &= 2h^2 P^T (\mathrm{d}p_1^T p_1 + \alpha \mathrm{d}p_2^T p_2) \in \mathbb{R}^{2(n+1)^2 \times 1}, \\ \hat{H}_1 &= h^2 P^T (\mathrm{d}p_1^T \mathrm{d}p_1 + \alpha \mathrm{d}p_2^T \mathrm{d}p_2) P \in \mathbb{R}^{2(n+1)^2 \times 2(n+1)^2}. \end{cases} \tag{19}$$

where $\mathrm{d}p_1 = [\Lambda D_1 - \mathrm{diag}(D_1\mathbf{T}./t)\Gamma; \Lambda D_2 - \mathrm{diag}(D_2\mathbf{T}./t)\Gamma]$, $\mathrm{d}p_2 = \sum_{i=1}^{2} \mathrm{diag}(D_i\mathbf{T}./\mathrm{LT} - D_i(\mathbf{T}+\mathbf{R})./\mathrm{LTR})D_i$, $\Lambda = \mathrm{diag}(1./\mathrm{LT})$, $t = \mathrm{LT}^{.3}$, $\Gamma = \sum_{i=1}^{2} \mathrm{diag}(D_i\mathbf{T})D_i$ and $\mathrm{diag}(v)$ is a diagonal matrix with $v$ on its main diagonal.

**Remark 2** Evaluating the deformed template image $\mathbf{T}$ must involve interpolation because $PX + PU$ are not in general pixel points. Here in our implementation, we choose B-splines for the interpolation.

For the discretized diffusion regularizer $\frac{\beta_1 h^2}{2} U^T A_1^T A_1 U + \frac{\beta_2 h^2}{2} U^T A_2 U$, its gradient and Hessian are respectively

$$\begin{cases} d_2 &= h^2(\beta_1 A_1^T A_1 + \beta_2 A_2)U \in \mathbb{R}^{2(n+1)^2 \times 1}, \\ H_2 &= h^2(\beta_1 A_1^T A_1 + \beta_2 A_2) \in \mathbb{R}^{2(n+1)^2 \times 2(n+1)^2}. \end{cases} \tag{20}$$

Finally, for the discretized Beltrami term $\frac{\beta h^2}{4} \phi(\mathbf{r}(U))e^T$, the gradient and approximated Hessian are as follows:

$$\begin{cases} d_3 &= \frac{\beta h^2}{4} \mathbf{dr}^T \mathrm{d}\phi(\mathbf{r}) \in \mathbb{R}^{2(n+1)^2 \times 1}, \\ \hat{H}_3 &= \frac{\beta h^2}{4} \mathbf{dr}^T \mathrm{d}^2\phi(\mathbf{r}) \mathbf{dr}. \end{cases} \tag{21}$$

where $\mathrm{d}\phi(\mathbf{r}) = (\phi'(\mathbf{r}_1), \ldots, \phi'(\mathbf{r}_{4n^2}))^T$ is the vector of derivatives of $\phi$ at all cell centers,

$$\begin{cases} \mathbf{dr} &= \mathrm{diag}(\mathbf{r}^1)\mathbf{dr}^2 + \mathrm{diag}(\mathbf{r}^2)\mathbf{dr}^1, \\ \mathbf{dr}^1 &= 2\mathrm{diag}(A_{31}U)A_{31} + 2\mathrm{diag}(A_{32}U)A_{32}, \\ \mathbf{dr}^2 &= -\mathrm{diag}(\mathbf{r}^2 \odot \mathbf{r}^2)[2\mathrm{diag}(A_{33}U + 2)A_{33} + 2\mathrm{diag}(A_{34}U)A_{34}], \end{cases} \tag{22}$$

$\odot$ denotes a Hadamard product, $\mathbf{dr}$, $\mathbf{dr}^1$, $\mathbf{dr}^2$ are the Jacobian of $\mathbf{r}$, $\mathbf{r}^1$, $\mathbf{r}^2$ with respect to $U$ respectively, $[\mathrm{d}\phi(\mathbf{r})]_\ell$ is the $\ell$th component of $\mathrm{d}\phi(\mathbf{r})$ and $\mathrm{d}^2\phi(\mathbf{r})$ is the Hessian of $\phi$ with respect to $\mathbf{r}$, which is a diagonal matrix whose $i$th diagonal element is $\phi''(\mathbf{r}_i)$, $1 \le i \le 4n^2$. More details about $\mathbf{r}^1$, $\mathbf{r}^2$, $A_{31}$, $A_{32}$, $A_{33}$ and $A_{34}$ are shown in Appendix 1.

Therefore, combining the above results for 3 terms, we can obtain the gradient

$$d_J = d_1 + d_2 + d_3 \tag{23}$$

and the approximated Hessian of (17):

$$H = \hat{H}_1 + H_2 + \hat{H}_3. \tag{24}$$

### 4.2.2 Search Direction

With the above approximated Hessian (24), in each outer (nonlinear) iteration, we solve the Gauss-Newton system

$$H\delta U = -d_J \tag{25}$$

to obtain the search direction $\delta U$ for (17). Because $H$ is symmetric positive semi-definite, in our implementation, we choose MINRES with diagonal preconditioning as the numerical solver [4, 36].

### 4.2.3 Step Length

Here, we choose a popular inexact line search condition, *Armijo condition*, which determines a step length $\theta$ that satisfies the following sufficient decrease condition:

$$\mathcal{J}(U + \theta\delta U) < \mathcal{J}(U) + \theta\eta {d_{\mathcal{J}}}^T \delta U. \tag{26}$$

Here, we set $\eta = 10^{-4}$ and use the backtracking approach to find a suitable $\theta$. In addition, we need to check that $\mathbf{r}(U)$ is smaller than 1 which is the norm of the discretized Beltrami coefficient. For more details, please refer to [25, 34, 41].

### 4.2.4 Stopping Criteria

In the implementation, we choose the stopping criteria used in [33]:

(1.a) $\|J(U^{i+1}) - J(U^i)\| \le \tau_J(1 + \|J(U^0)\|)$,
(1.b) $\|U^{i+1} - U^i\| \le \tau_W(1 + \|X + U^0\|)$,
(1.c) $\|d_J\| \le \tau_G(1 + \|J(U^0)\|)$,
(2) $\|d_J\| \le$ eps,
(3) $i \ge$ MaxIter.

Here, eps is the machine precision and MaxIter is the maximal number of outer iterations. We set $\tau_J = 10^{-3}$, $\tau_W = 10^{-2}$ and $\tau_G = 10^{-2}$. If any one of (1) (2) and (3) is satisfied, the iterations are terminated. Hence, a Gauss-Newton numerical scheme with Armijo line search can be developed and summarized in Algorithm 1.

**Algorithm 1** Gauss-Newton scheme by using Armijo line search for Image Registration: $U \leftarrow \text{GNAIR}(\alpha, \beta_1, \beta_2, \gamma, U^0, T, R)$

Step 1: Set $i = 0$ at the solution point $U^i = U^0$.
Step 2: For (17), compute the energy functional $J(U^i)$, its gradient $d_J^i$ and the approximated Hessian $H^i$ by (24).
**while** "none of the listed 3 stopping criteria are satisfied" **do**
  Solve the Gauss-Newton equation: $H^i \delta U^i = -d_J^i$;
  Use Line Search to find step length $\theta$;
  $U^{i+1} = U^i + \theta \delta U^i$;
  $i = i + 1$;
  Compute $J(U^i)$, $d_J^i$ and $H^i$;
**end while**

### 4.2.5 Multi-level Strategy

A multi-level strategy is a standard technique in image registration. In the multi-level strategy, we firstly coarsen the template $T$ and the reference $R$ by $L$ levels. Then we can obtain $U_1$ by solving our model (6) on the coarsest level. In order to give a good initial guess for the finer level, we adopt an interpolation operator on $U_1$ to obtain $U_2^0$ as the initial guess for the next level. We repeat this process and can get the final registration on the finest level. The most important advantage of the multi-level strategy is that it can save computation time because of less variables on the coarser level than on the fine level. In addition, it can help to avoid trapping into a local minimum.

### 4.2.6 Convergence Result

Our above described Algorithm 1 will converge to a stationary point of our new model. Details are shown in Theorem 1 of Appendix 2 below.

## 5 Numerical Results

In this section, we will show some numerical results to illustrate the performances of our proposed model (6) using Gauss-Newton method called **GNR**. We compare with the standard **NGF** [32] and the Augmented Lagrangian approach for solving a similar model [43] called **ALMR**, which uses the same regularization and fitting terms. However, the local invertibility of the map is guaranteed by imposing an inequality constraint on the model. For more details about the augmented Lagrangian method, we refer to [3, 38, 42] and the reference therein.

**ALMR**. Alternating iteration is another popular method which might be applied to (6). However, below, we shall consider it for a related model [43] that uses a constrained optimization (different from (6)):

$$\begin{cases} \min_{u \in \mathcal{H}} \{\mathcal{J}_1(u) = S(u) + \frac{\lambda}{2} D^{GF}(u) + \frac{\lambda}{2} D^{TM}(u)\}, \\ \text{w.r.t} \quad \mathcal{C}_\epsilon(u) = \det(I + \nabla u) \geq \epsilon, \end{cases} \tag{27}$$

where imposing the constraint is a competing way of ensuring a diffeomorphic transformation.

To reformulate (27), introducing variables $K$, $\mathbf{p}$ and $\mathbf{n}$, we solve the following constrained minimization problem:

$$\begin{cases} \min_{u,K,\mathbf{p},\mathbf{n}} \{S(u) + \frac{\lambda}{2} \int_\Omega (\mathbf{n} - \nabla_n R)^2 \mathrm{d}x + \frac{\lambda}{2} \int_\Omega (|\mathbf{p}| + |\nabla R| - |\mathbf{m}|)^2 \, \mathrm{d}x\}, \\ \text{w.r.t} \quad K = T(x + u), \quad \mathbf{p} = \nabla K, \quad |\mathbf{p}|\mathbf{n} = \mathbf{p}, \quad \mathbf{m} = \mathbf{p} + \nabla R, \quad \mathcal{C} > 0. \end{cases} \tag{28}$$

Then, the augmented Lagrangian functional corresponding to the constrained optimization problem (28) is defined as follows:

$$\begin{aligned} &\mathcal{L}_1(u, K, \mathbf{p}, \mathbf{n}, \mathbf{m}, \lambda_1, \lambda_2, \lambda_3, \lambda_4, \lambda_5) \\ &= S(u) + \frac{\lambda}{2} \int_\Omega (\mathbf{n} - \nabla_n R)^2 \mathrm{d}x + \frac{\lambda}{2} \int_\Omega (|\mathbf{p}| + |\nabla R| - |\mathbf{m}|)^2 \, \mathrm{d}x \\ &+ \frac{r_2}{2} \int_\Omega (\mathbf{p} - \nabla K)^2 \mathrm{d}x + \frac{r_3}{2} \int_\Omega (\mathbf{p} - |\mathbf{p}|\mathbf{n})^2 \mathrm{d}x + \frac{r_4}{2} \int_\Omega (\mathbf{p} + \nabla R - \mathbf{m})^2 \mathrm{d}x \\ &+ \int_\Omega (T(x + u) - K)\lambda_1 \mathrm{d}x + \int_\Omega (\mathbf{p} - \nabla K) \cdot \lambda_2 \mathrm{d}x + \int_\Omega (\mathbf{p} - |\mathbf{p}|\mathbf{n}) \cdot \lambda_3 \mathrm{d}x \\ &+ \int_\Omega (\mathbf{p} + \nabla R - \mathbf{m}) \cdot \lambda_4 \, \mathrm{d}x + \frac{r_1}{2} \int_\Omega (T(x + u) - K)^2 \mathrm{d}x + \frac{1}{2\sigma} \int_\Omega \mathcal{C}_s(u, \lambda_5) \, \mathrm{d}x, \end{aligned} \tag{29}$$

where

$$\mathcal{C}_s(u, \lambda_5) = [\min\{0, \sigma(\mathcal{C}(u) - \epsilon) - \lambda_5\}]^2 - \lambda_5^2, \tag{30}$$

$\epsilon > 0$ is a small parameter, $\sigma > 0$ and $\lambda_i (i = 1, \ldots, 5)$ are the Lagrange multipliers. The augmented Lagrangian algorithm is shown in Algorithm 2.

**Algorithm 2** Augmented Lagrangian method

1. Initialization: $\boldsymbol{u}^0$, $K^0$, $\mathbf{p}^0$, $\mathbf{n}^0$,$\mathbf{m}^0$ and $\lambda_1^0$, $\lambda_2^0$, $\lambda_3^0$, $\lambda_4^0$ and $\lambda_5^0$.
2. Iterate for $k = 1, 2, \ldots$ until a required tolerance:
   — compute an approximate minimizers $\mathbf{u}^{k+1}$, $K^{k+1}$, $\mathbf{p}^{k+1}$, $\mathbf{n}^{k+1}$ and $\mathbf{m}^{k+1}$ of the augmented Lagrangian functional with the fixed Lagrange multipliers $\lambda_1^k$, $\lambda_2^k$, $\lambda_3^k$, $\lambda_4^k$ and $\lambda_5^k$:

$$\left[\boldsymbol{u}^{k+1}, K^{k+1}, \mathbf{p}^{k+1}, \mathbf{n}^{k+1}, \mathbf{m}^{k+1}\right] = \operatorname*{arg\,min}_{\boldsymbol{u}, K, \mathbf{p}, \mathbf{n}} \mathcal{L}_1(\boldsymbol{u}, K, \mathbf{p}, \mathbf{n}, \mathbf{m}, \lambda_1^k, \lambda_2^k, \lambda_3^k, \lambda_4^k, \lambda_5^k). \tag{31}$$

   — Update Lagrange multipliers

$$\lambda_1^{k+1} = \lambda_1^k + r_1(T(\boldsymbol{x} + \boldsymbol{u}^{k+1}) - K^{k+1}), \tag{32}$$
$$\lambda_2^{k+1} = \lambda_2^k + r_2(\mathbf{p}^{k+1} - \nabla K^{k+1}), \tag{33}$$
$$\lambda_3^{k+1} = \lambda_3^k + r_3(\mathbf{p}^{k+1} - |\mathbf{p}^{k+1}|\mathbf{n}^{k+1}), \tag{34}$$
$$\lambda_4^{k+1} = \lambda_4^k + r_4(\mathbf{m}^{k+1} - \mathbf{p}^{k+1} - \nabla R), \tag{35}$$
$$\lambda_5^{k+1} = \max\{0, \lambda_5^k - \sigma \mathcal{C}_\epsilon(\boldsymbol{u}^{k+1})\}, \tag{36}$$

In practice, the minimization problem (29) or (31) is decomposed into a number of sub-problems, each of which can be solved quickly. However, the convergence of the augmented Lagrangian iterations for this case is not guaranteed due to the non-convexity of overall registration problem. Currently this is a major weakness of **ALMR** while the convergence of **GNR** (even if a bit slower) can be proved and hence recommended.

In order to reduce the number of parameters to tune, we set $\lambda = 15$, $\beta_1 = 0.005$, $\beta_2 = 0.1 \times \beta_1$ $r_1 = 5$, $r_2 = 10$ and $r_3 = r_4 = 100$ in all numerical experiments unless stated otherwise. We consider $N_{max} = 70$ as the maximum number of iterations for **ALMR** from Algorithm 2 and we stop the iterations before reaching $N_{max} = 70$ if the following stopping criterion

$$\frac{\|\mathbf{p}^k + \nabla R - \mathbf{m}^k\|_{L^1}}{\sqrt{l \times c}} \le \tau$$

is satisfied for a given tolerance $\tau = 10^{-3}$, where $l$ and $c$ are the numbers of rows and columns in the image

For all compared methods, we set the zero vector as the initial guess $U^0$. To measure the quality of the registered images, we use the following quantities

$$\text{GFer} = \frac{D^{GF}(\boldsymbol{u})}{D^{GF}(\boldsymbol{u}^0)}, \tag{37}$$

$$\text{NGFer} = \frac{D^{NGF}(\boldsymbol{u})}{D^{NGF}(\boldsymbol{u}^0)}, \tag{38}$$

and

$$\text{MIer} = -D^{MI}(\boldsymbol{u}). \tag{39}$$

The good result means that it can lead to small GFer, small NGFer and large MIer. All the codes are implemented by Matlab R2019b on a PC with 3.4 GHz Intel(R) Core(TM) i5-3570 processor and 12 GB RAM.

## 5.1 *Example 1*

In this example, we consider a pair of images displayed in Fig. 2a, b. The resolution is $256 \times 256$. In order to choose the parameter easily, in this example, we fix $\alpha$ and set $\alpha = 0.01$.

Firstly, we consider the model without Beltrami control term, namely $\gamma = 0$. For the parameters of regularizers, we set two pairs $(\beta_1, \beta_2) = (50, 2)$ and $(\beta_1, \beta_2) = (50, 5)$. The corresponding deformed templates and transformations are shown in Fig. 2d, e, g, h. From Fig. 2f, i, we can find that the deformed templates generated by these two pairs of parameters are visually satisfied. In addition, these two choices give similar measurements: GFer = 0.82, NGFer = 0.81, MIer = 0.58 and GFer = 0.83, NGFer = 0.84, MIer = 0.57 respectively. However, the first choice leads to a transformation containing folding because the minimum of the Jacobian determinant of the transformation is negative but the second choice produces a smooth transformation without folding because the minimum of the Jacobian determinant of the transformation is positive.

Since first and second order regularizers just control the smoothness, in order to overcome this drawback, we keep $(\beta_1, \beta_2) = (50, 2)$ unchanged and choose a suitable $\gamma$. Here, we set $\gamma = 10$. Figure 3a, b shows the corresponding deformed template and transformation. From Fig. 3c, the deformed template is similar visually with the previous one without controlling the Beltrami coefficient and the measurements are also similar (GFer = 0.82, NGFer = 0.82 and MIer = 0.57). But the minimum of the Jacobian determinant of the transformation is positive, which illustrates that the transformation is diffeomorphic. In the same figure, we also give the result of **ALMR** model, which shows again from the overlay of $T(\varphi)$ and the reference $R$ that the template image $T$ is well registered to $R$.

Now, we investigate the sensitivity of $\gamma$. From Table 1, we can find that when we fix $\alpha$, $\beta_1$ and $\beta_2$ and change $\gamma$, GFer, NGFer and MIer are robust and at the same time, the minimum of the Jacobian determinant of the transformations are all positive. This indicates that the Beltrami control term is not sensitive.

In addition, we also investigate the convergence of the algorithm for our model. Here, we force the relative norm of the gradient of the approximated solution to reach $10^{-3}$ although it only runs several iterations by using the practical stopping criteria. Here, according to Fig. 4, we can find that the algorithm for our model is convergent.

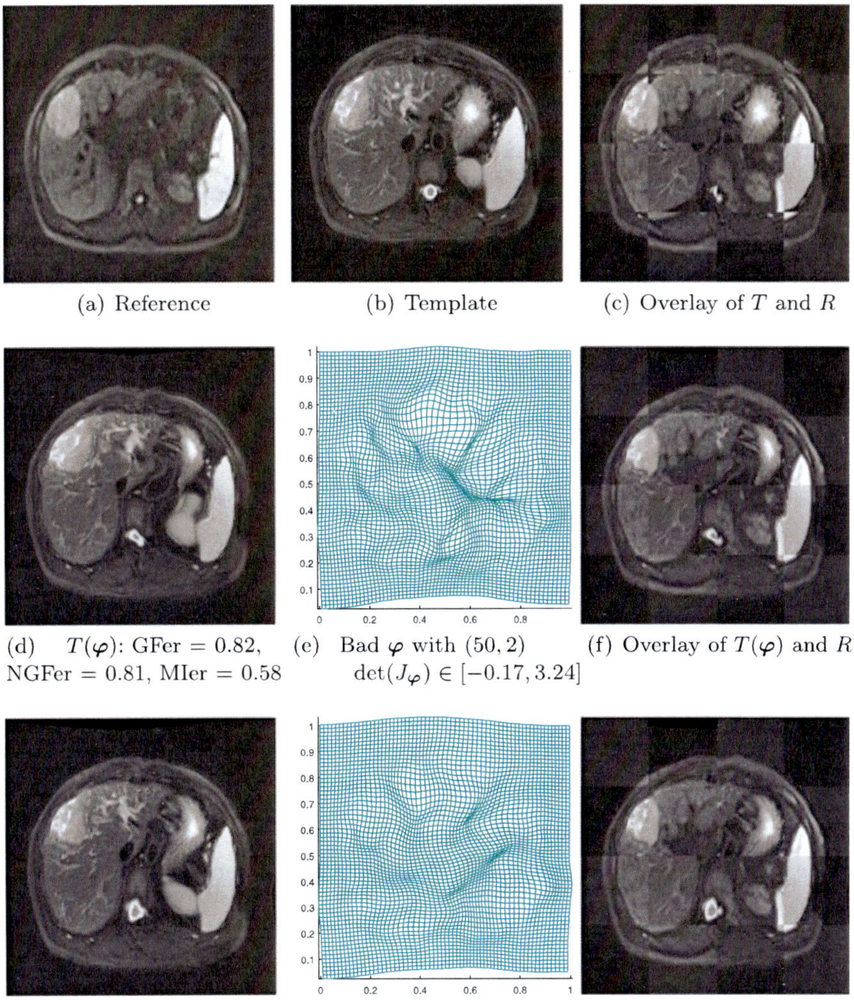

(a) Reference (b) Template (c) Overlay of $T$ and $R$

(d) $T(\varphi)$: GFer = 0.82, NGFer = 0.81, MIer = 0.58 (e) Bad $\varphi$ with $(50, 2)$ $\det(J_\varphi) \in [-0.17, 3.24]$ (f) Overlay of $T(\varphi)$ and $R$

(g) $T(\varphi)$: GFer = 0.83, , NGFer = 0.84, MIer = 0.57 (h) Good $\varphi$ with $(50, 5)$ $\det(J_\varphi) \in [0.17, 2.30]$ (i) Overlay of $T(\varphi)$ and $R$

**Fig. 2** Example 1 **without the Beltrami control term**: the first row shows the reference, template and overlay of the reference and template. The second and third rows show the deformed templates and transformations obtained by two pairs of parameters $(\beta_1, \beta_2) = (50, 2)$ and $(\beta_1, \beta_2) = (50, 5)$, respectively. The results are visually similar but the transformations are not both one-to-one. The first choice leads to a mesh with folding because the minimum of the Jacobian determinant of the transformation is negative

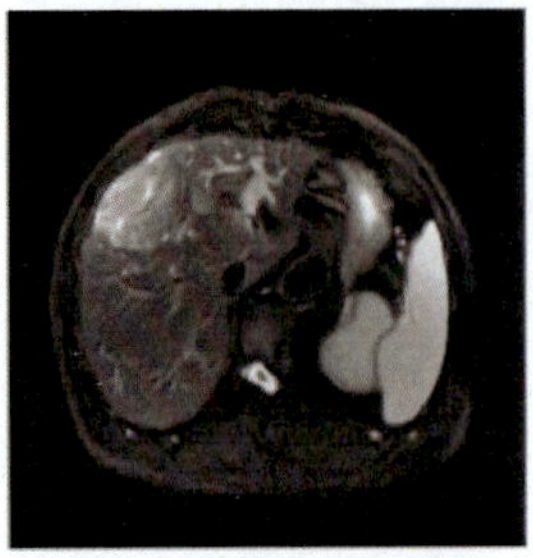

(a) $T(\varphi)$, GFer = 0.82, NGFer = 0.82, MIer = 0.57

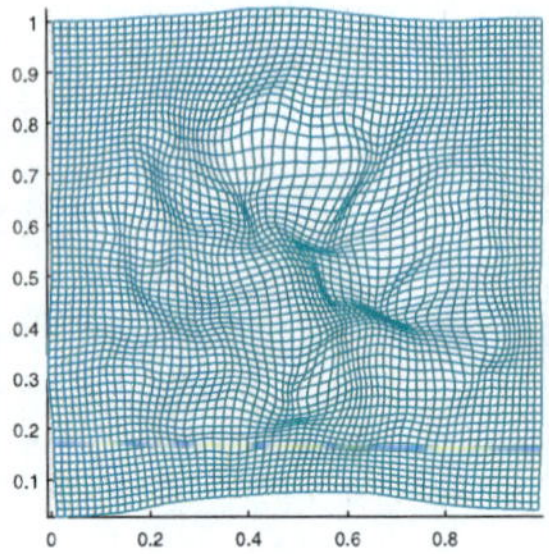


(b) **GNR** $\varphi$: $\det(J_\varphi) \in [0.05, 3.48]$

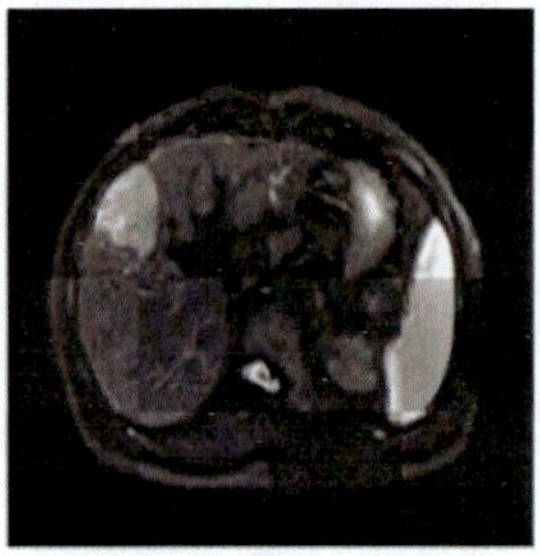

(c) **GNR**: $T(\varphi)$ over $R$

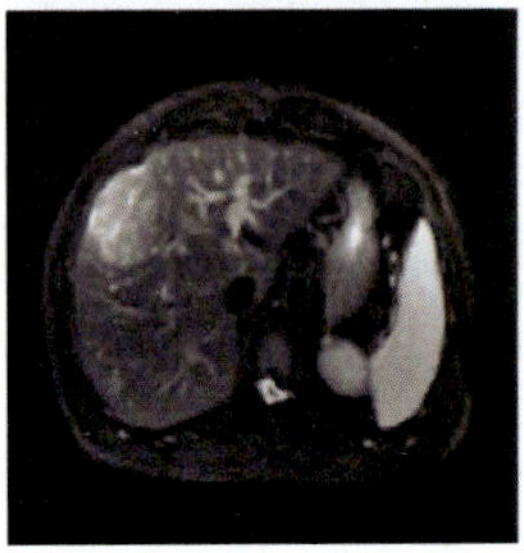

(d) $T(\varphi)$, GFer = 0.97, NGFer = 0.88, MIer = 0.59

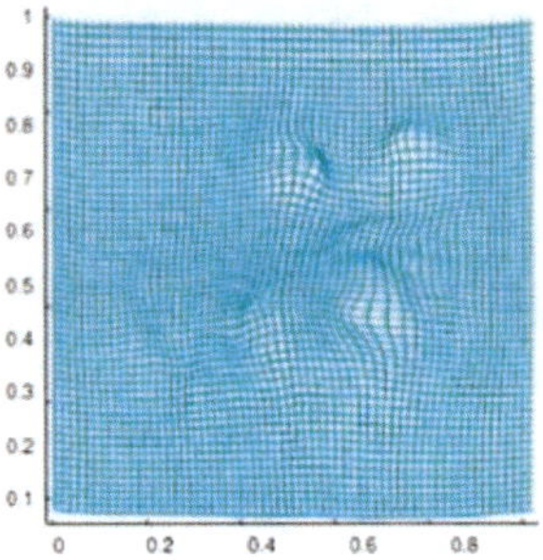


(e) **ALMR** $\varphi$: $\det(J_\varphi) \in [0.04, 3.09]$

(f) **ALMR**: $T(\varphi)$ over $R$

**Fig. 3** Example 1: the deformed template and transformation are generated by $(\beta_1, \beta_2, \gamma) = (50, 2, 10)$. The results are visually satisfied and the transformation is one-to-one. Second row: the deformed template obtained by **ALMR** and its overlay with the reference $R$

**Table 1** Example 1: measurements obtained by using $\alpha = 10^{-2}$, $\beta_1 = 50$ and $\beta_2 = 2$

| $\gamma$ | GFer | NGFer | MIer | min $\det(J_\varphi)$ | max $\det(J_\varphi)$ |
|---|---|---|---|---|---|
| 1 | 0.82 | 0.82 | 0.57 | 0.01 | 3.14 |
| 10 | 0.82 | 0.82 | 0.57 | 0.05 | 3.48 |
| 100 | 0.82 | 0.82 | 0.57 | 0.06 | 3.13 |
| 1000 | 0.82 | 0.82 | 0.57 | 0.21 | 3.11 |

Hence, this example illustrates that our new control term can effectively control the transformation and lead to an accurate registration. Meanwhile, the new control term can make this model more robust.

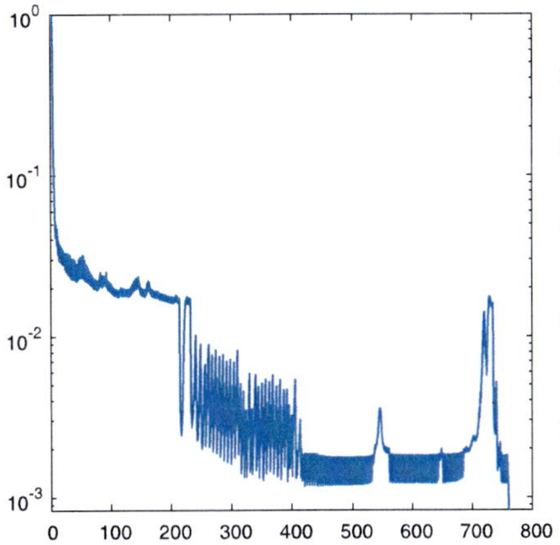


(a) Relative norm of the gradient

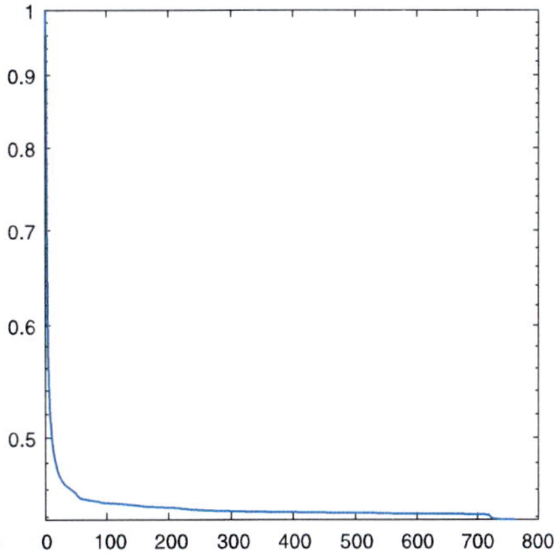


(b) Relative norm of the function value

**Fig. 4** Example 1: Relative norm of the gradient and relative norm of the function value by the parameter $(\alpha, \beta_1, \beta_2, \gamma) = (0.01, 50, 2, 10)$. Here, we can notice that our algorithm is convergent

## *5.2 Example 2*

In this example, we consider another pair of $256 \times 256$ images (Fig. 5a, b). Again, in order to reduce the complexity of choosing parameters, we fix $\alpha = 10^{-1}$ in this example.

Firstly, we set $\beta_1 = 50, \beta_2 = 10$ and $\gamma = 0$. From Fig. 5d–f, although the deformed template is satisfied visually, we can find that the resulting transformation has folding since the minimum of the Jacobian determinant is negative.

As a comparison, we also test the model of the standard **NGF** [32] with the same first- and second-order regularizer. Here, we test three pairs of $(\beta_1, \beta_2)$ and the corresponding results are shown in Fig. 6. We can find that for the fitting term, if we choose **NGF**, it is very hard to choose the suitable parameters to get a good registration, namely, simultaneously get a diffeomorphic transformation and a visually satisfied deformed template. In order to overcome this difficulty, we keep $\beta_1$, $\beta_2$ unchanged and choose $\gamma$ as 1, 10 and 100 separately. Figure 7 shows that they can all generate visually satisfied deformed template and diffeomorphic transformations. Specifically, according to Fig. 7, we can see that the measurements obtained by these choices are very similar, which again demonstrates that this model can be more robust through combining the Beltrami control term. We also give the result of **ALMR** model in Fig. 8. We can observe from overlay of the registered and the reference images that all models work fine in producing acceptable registration result.

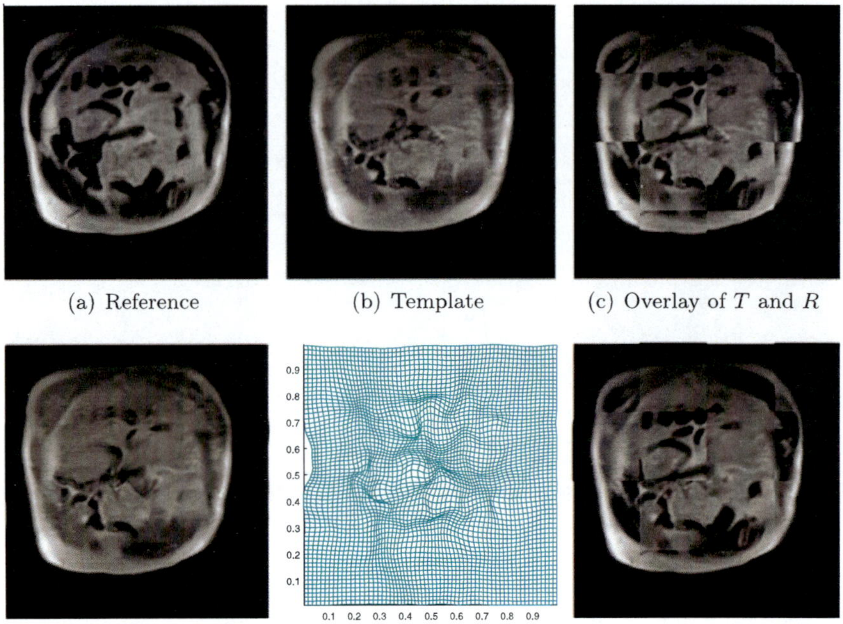


(a) Reference (b) Template (c) Overlay of $T$ and $R$

(d) $T(\varphi)$, GFer $= 0.62$, NGFer $= 0.68$, MIer $= 0.83$ (e) **GNR**: $\varphi$, $\det(J_\varphi) \in [-0.19, 3.13]$ (f) **GNR**: $T(\varphi)$ over $R$

**Fig. 5** Example 2 by the new model **GNR** without using the control term $C$: the resulting transformation is not diffeomorphic although the deformed template is visually satisfied

In summary, when the **ALMR**, the **NGF** and the **GNR** work, the latter has the largest MIer similarity (indicating better quality). However, **NGF** (or taking out an extra control term for **ALMR** and **GNR**) can fail to deliver a valid result (with negative $\det \nabla \boldsymbol{y}$) if the parameters are not chosen correctly. Although **ALMR** is completive to **GNR** (and takes less time to converge in practice), only the convergence of **GNR** can be proved. Hence our model **GNR** is robust and can be recommended for multi-modal registration.

## 6 Conclusions

Image registration is an increasingly important and often challenging image processing task. The quality of the transformation requires suitable control. In this Chapter to improve a multi-modality registration model, we propose a novel term motivated by Beltrami coefficient, which can lead to a diffeomorphic transformation. The advantage of the term lies in no bias imposed on its Jacobian of the transformation's determinant. By employing first-discretize-then-optimize method, we design an effective

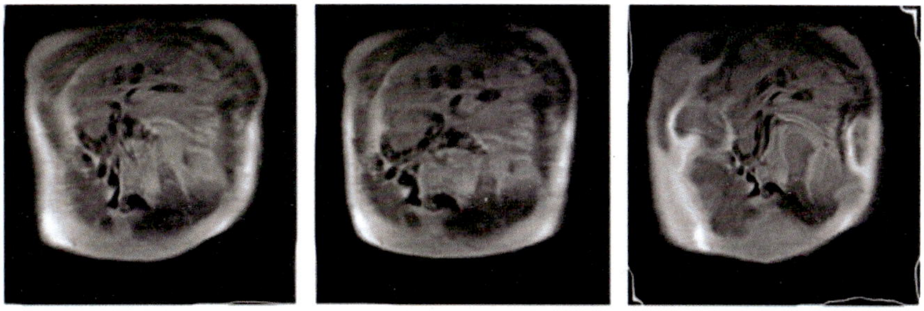

(a) $T(\varphi)$ with GFer $= 1.08$, NGFer $= 0.96$, MIer $= 0.50$ with $(\beta_1, \beta_2) = (0.1, 0.001)$ (b) $T(\varphi)$ with GFer $= 1.07$, NGFer $= 0.96$, MIer $= 0.53$ with $(\beta_1, \beta_2) = (0.01, 0.01)$ (c) $T(\varphi)$: GFer $= 1.14$, NGFer $= 0.99$, MIer $= 0.50$ with $(\beta_1, \beta_2) = (0.01, 10^{-4})$

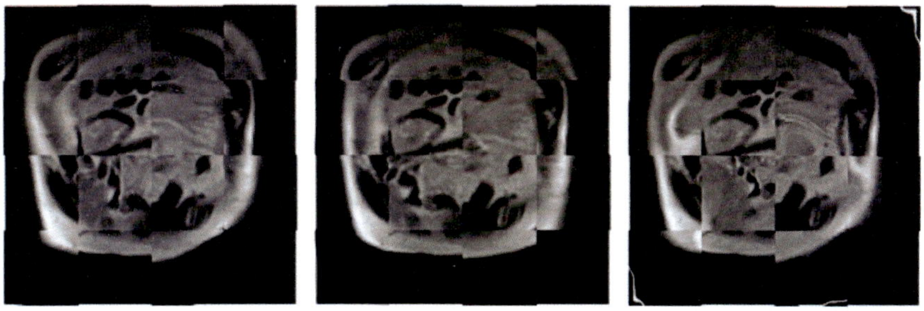

(d) $T(\varphi)$ over $R$ (e) $T(\varphi)$ over $R$ (f) $T(\varphi)$ over $R$

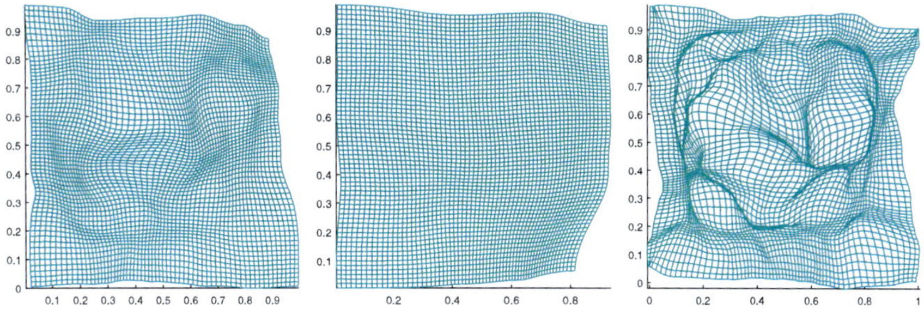


(g) $\varphi$ $\det(J_\varphi) \in [0.27, 1.94]$ (h) $\varphi$ $\det(J_\varphi) \in [0.54, 1.19]$ (i) $\varphi$ $\det(J_\varphi) \in [-0.89, 4.10]$

**Fig. 6** Example 2 by the **GNR without** imposing a control term. Each column shows results of a different choice of $(\beta_1, \beta_2)$ balancing first- and second-order regularizers: the deformed template, overlay of $T(\varphi)$ and $R$, and the transformation. Clearly the last column obtains the incorrect $\varphi$

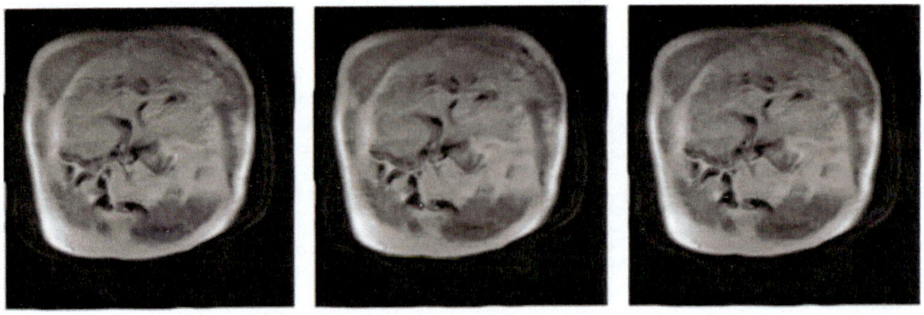

(a) $T(\varphi)$: GFer = 0.62, NGFer = 0.68, MIer = 0.83 with $\gamma = 1$ (b) $T(\varphi)$: GFer = 0.63, NGFer = 0.68, MIer = 0.83 with $\gamma = 10$ (c) $T(\varphi)$: GFer = 0.63, NGFer = 0.68, MIer = 0.83 with $\gamma = 100$

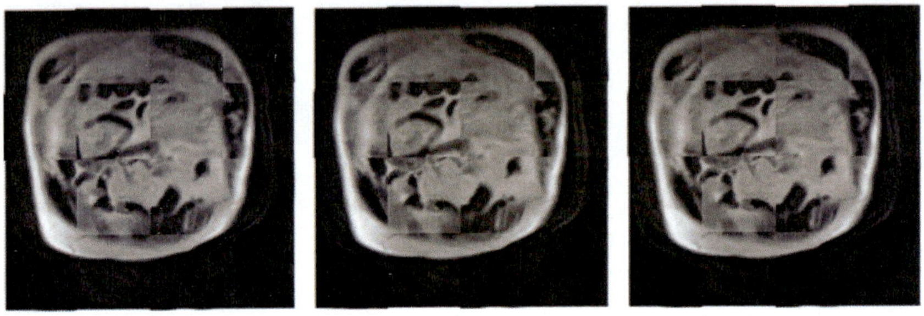

(d) $T(\varphi)$ over $R$ (e) $T(\varphi)$ over $R$ (f) $T(\varphi)$ over $R$

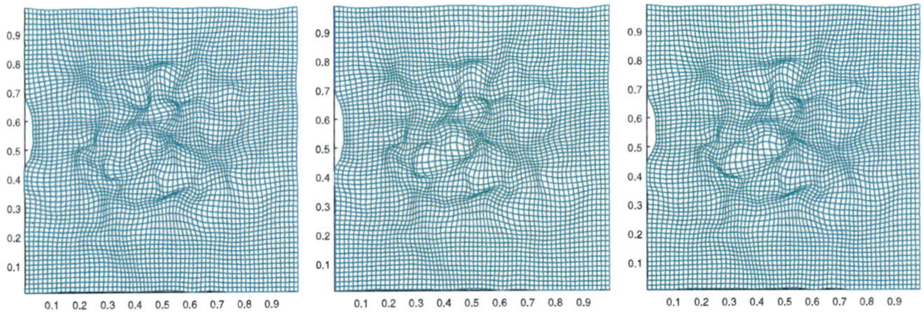

(g) $\varphi$: $\det(J_\varphi) \in [0.01, 2.54]$ (h) $\varphi$: $\det(J_\varphi) \in [0.01, 3.27]$ (i) $\varphi$: $\det(J_\varphi) \in [0.02, 3.19]$

**Fig. 7** Example 2 by the new model **GNR**. By using the control term for each choice of $\gamma$ (by column), the resulting transformation is diffeomorphic and the deformed template is also visually pleasing

solver to solve our proposed model numerically. Experimental tests confirm that our proposed model performs well in multi-modality images registration. In addition, with the help of the Beltrami control term, the proposed model is more robust with respect to the parameters. Future work will investigate extension of this work to a deep learning framework [45].

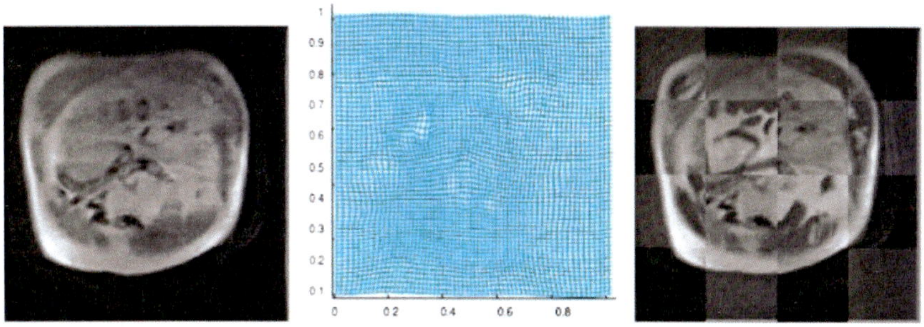


(a) $T(\varphi)$, GFer = 0.51, NGFer = 0.62, MIer = 0.78 (b) $\varphi$: $\det(J_\varphi) \in [0.11, 3.15]$ (c) **ALMR**: $T(\varphi)$ over $R$

**Fig. 8** Example 2 by **ALMR** model. The deformed template is also visually close to the reference R

## Appendix 1—Computation of the Vector $\mathbf{r}(U)$

First of all, denote the 3 vertices of this triangle by $V_1 = \boldsymbol{x}^{1,1}$, $V_2 = \boldsymbol{x}^{2,1}$ and $V_5 = \boldsymbol{x}^{1.5,1.5}$ in Fig. 1. Set $\mathbf{L}(V_1) = (u_1^{1,1}, u_2^{1,1})$, $\mathbf{L}(V_2) = (u_1^{2,1}, u_2^{2,1})$ at the vertex pixels, and $\mathbf{L}(V_5) = (u_1^{1.5,1.5}, u_2^{1.5,1.5})$ at the cell centre (approximated values). Here the linear approximations are $\mathbf{L}(x_1, x_2) = (a_1x_1 + a_2x_2 + a_3, a_4x_1 + a_5x_2 + a_6)$.

After substituting $V_1$, $V_2$ and $V_5$ into $\mathbf{L}$, we get

$$\begin{pmatrix} a_1 \\ a_2 \end{pmatrix} = \frac{1}{\det} \begin{pmatrix} x_2^1 - x_2^{1.5} & -x_2^1 + x_2^{1.5} \\ -x_1^2 + x_1^{1.5} & x_1^1 - x_1^{1.5} \end{pmatrix} \begin{pmatrix} u_1^{1,1} - u_1^{1.5,1.5} \\ u_1^{2,1} - u_1^{1.5,1.5} \end{pmatrix}, \tag{40}$$

$$\begin{pmatrix} a_4 \\ a_5 \end{pmatrix} = \frac{1}{\det} \begin{pmatrix} x_2^1 - x_2^{1.5} & -x_2^1 + x_2^{1.5} \\ -x_1^2 + x_1^{1.5} & x_1^1 - x_1^{1.5} \end{pmatrix} \begin{pmatrix} u_2^{1,1} - u_2^{1.5,1.5} \\ u_2^{2,1} - u_2^{1.5,1.5} \end{pmatrix}, \tag{41}$$

where $\det = \begin{vmatrix} x_1^1 - x_1^{1.5} & x_2^1 - x_2^{1.5} \\ x_1^2 - x_1^{1.5} & x_2^1 - x_2^{1.5} \end{vmatrix}$.

According to (40) and (41), we can formulate two matrices $D1 \in \mathbb{R}^{4n^2 \times (n+1)^2}$ and $D2 \in \mathbb{R}^{4n^2 \times (n+1)^2}$ such that

$$A_{31} = [D1, -D2], A_{32} = [D2, D1], A_{33} = [D1, D2], A_{34} = [D2, -D1].$$

Then using the Hadamard product $\odot$, we get a compact form for

$$\begin{cases} \mathbf{r}^1(U) = A_{31}U \odot A_{31}U + A_{32}U \odot A_{32}U, \\ \mathbf{r}^2(U) = 1./((A_{33}U + 2) \odot (A_{33}U + 2) + A_{34}U \odot A_{34}U), \\ \mathbf{r}(U) \;\; = \mathbf{r}^1 \odot \mathbf{r}^2 \quad \in \mathbb{R}^{4n^2 \times 1}. \end{cases} \tag{42}$$

## Appendix 2—The Global Convergence of Algorithm 1

In order to discuss the global convergence result of Algorithm 1 for the discretized optimization problem (17), we first review two lemmas.

**Lemma 1** ([25]) *For the unconstrained optimization problem*

$$\min_U J(U)$$

*let an iterative sequence be defined by* $U^{i+1} = U^i + \theta \delta U^i$*, where* $\delta U^i = -(H^i)^{-1} d_J(U^i)$ *and* $\theta$ *is obtained by Armijo condition. Assume that three conditions are met: (i).* $d_J$ *be Lipschitz continuous; (ii). the matrices* $H^i$ *are SPD (iii). there exist constants* $\bar{\kappa}$ *and* $M$ *such that the condition number* $\kappa(H^i) \le \bar{\kappa}$ *and the norm* $||H^i|| \le M$ *for all* $i$*. Then either* $J(U^i)$ *is unbounded from below or*

$$\lim_{i\to\infty} d_J(U^i) = 0 \tag{43}$$

*and hence any limit point of the sequence of iterates is a stationary point.*

**Lemma 2** *Let a matrix be comprised of* 3 *submatrices* $H = H_1 + H_2 + H_3$*. If* $H_1$ *and* $H_2$ *are symmetric positive semi-definite and* $H_3$ *is SPD, then* $H$ *is SPD with* $\lambda_{H_3} \le \lambda_H$*, where* $\lambda_{H_3}$ *and* $\lambda_H$ *are the minimum eigenvalues of* $H_3$ *and* $H$ *separately.*

***Proof*** According to Rayleigh quotient, we can find a vector $v$ such that

$$\lambda_H = \frac{v^T H v}{v^T v}. \tag{44}$$

Then we have

$$\lambda_{H_3} \le \frac{v^T H_1 v}{v^T v} + \frac{v^T H_2 v}{v^T v} + \frac{v^T H_3 v}{v^T v} = \frac{v^T H v}{v^T v} = \lambda_H, \tag{45}$$

which completes the proof.

In the above discretization leading to (17), we do not need to introduce the boundary condition. However for theory purpose, in the following, we will prove our convergence result under the Dirichlet boundary condition (namely, the boundary is fixed) and this condition is needed to prove the symmetric positive definite (SPD) property of the approximated Hessian. In practical implementation, such a condition is not required as confirmed by experiments.

In addition, define an important set $\mathcal{X} := \{U \mid \mathbf{r}(U)_\ell \le 1 - \epsilon, 1 \le \ell \le 4n^2\}$ for small $\epsilon$. So $U \in \mathcal{X}$ means that the transformation is diffeomorphic. Under the suitable $\gamma$, each $U^i$ generated by Algorithm 1 is in the $\mathcal{X}$.

**Theorem 1** *Assume that $T$ and $R$ are twice continuously differentiable. For (17), by using Algorithm 1, we obtain*

$$\lim_{i\to\infty} d_J(U^i) = 0 \tag{46}$$

*and hence any limit point of the sequence of iterates produced by Algorithm 1 is a stationary point.*

***Proof*** It suffices to show that Algorithm 1 satisfies the requirements of Lemma 1. Recall $\mathbf{r}(U)$ and we can see that it is continuous. Here, we use the Dirichlet boundary condition and we can assume that $\|U\|$ is bounded. Then $\mathbf{r}(U)$ is a continuous mapping from a compact set to $\mathbb{R}^{4n^2\times 1}$ and $\mathbf{r}(U)$ is proper. So for some small $\epsilon > 0$, $\mathcal{X}$ is compact.

Firstly, we show that in $\mathcal{X}$, $d_J$ of (17) is Lipschitz continuous. The term $\phi(\mathbf{r}(U))e^T$ in the (17) is twice continuously differentiable with respect to $U \in \mathcal{X}$. In addition, $T$ and $R$ are twice continuously differentiable. So (17) is twice continuously differentiable with respect to $U \in \mathcal{X}$ and $d_J$ is Lipschitz continuous.

Secondly, we show that in $\mathcal{X}$, $H^i = \hat{H}_1^i + H_2^i + \hat{H}_3^i$ is SPD. By the construction of $\hat{H}_1^i$ and $\hat{H}_3^i$, they are symmetric positive semi-definite. $H_2^i$ is symmetric positive definite under the Dirichlet boundary condition. Consequently, according to Lemma 2, $H^i$ is SPD.

Thirdly, we show that both $\kappa(H^i)$ and $\|H^i\|$ are bounded. We notice that in each iteration, $H_2^i$ is constant and we can set $\|H_2^i\| = M_2$. For $\hat{H}_1^i$, we get its upper bound $M_1$ because $T$ is twice continuously differentiable and $\mathcal{X}$ is compact. $\phi$ is also twice continuously differentiable with respect to $U \in \mathcal{X}$, then we have $\|\hat{H}_3^i\| \le M_3$. Hence, we have

$$\|H^i\| \le \|\hat{H}_1^i\| + \|H_2^i\| + \|\hat{H}_3^i\| \le M_1 + M_2 + M_3. \tag{47}$$

So set $M = M_1 + M_2 + M_3$ and $\|H^i\| \le M$. Set $\sigma$ as the minimum eigenvalue of $H_2^i$. According to Lemma 2, the smallest eigenvalue $\lambda_{min}$ of $H^i$ should be larger than $\sigma$. The largest eigenvalue $\lambda_{max}$ of $H^i$ should be smaller than $M$ due to $\lambda_{max} \le \|H^i\|$. So the conditional number of $H^i$ is smaller than $\frac{M}{\sigma}$.

Finally, we can find that (17) has lower bound 0. Hence, by applying Lemma 1, we complete the proof.

## References


1. A. Angelov, M. Wagner, Multimodal image registration by elastic matching of edge sketches via optimal control. Journal of Industrial & Management Optimization **10**, 567–590 (2014)
2. N. Aronszajn, Theory of reproducing kernels. Transactions of the American mathematical society **68**, 337–404 (1950)
3. E. Bae, J. Shi, X.-C. Tai, Graph cuts for curvature based image denoising. IEEE Transactions on Image Processing **20**, 1199–1210 (2011)

4. R. Barrett, M.W. Berry, T.F. Chan, J. Demmel, J. Donato, J. Dongarra, V. Eijkhout, R. Pozo, C. Romine, H. Van der Vorst, Templates for the solution of linear systems: building blocks for iterative methods, vol. 43 (Siam, 1994)
5. A. Berlinet, C. Thomas-Agnan, *Reproducing Kernel Hilbert Spaces in Probability and Statistics* (Springer Science & Business Media, 2011)
6. K. Briechle, U.D. Hanebeck, Template matching using fast normalized cross correlation, in *Optical Pattern Recognition XII*, vol. 4387 (International Society for Optics and Photonics, 2001), pp. 95–102
7. M. Burger, J. Modersitzki, L. Ruthotto, A hyperelastic regularization energy for image registration. SIAM Journal on Scientific Computing **35**, B132–B148 (2013)
8. K. Chen, L. M. Lui, J. Modersitzki, Image and surface registration, in *Learning of Images, Shapes, and Forms*, ed. by R. Kimmel, X.-C. Tai, vol. 20 (Handbook of Numerical Analysis, North Holland - Elsevier, 2019), pp. 579–611
9. Y.M. Chen, J.L. Shi, M. Rao, J.S. Lee, Deformable multi-modal image registration by maximizing renyi's statistical dependence measure. Inverse Problems and Imaging **9**, 79–103 (2015)
10. N. Chumchob, Vectorial total variation-based regularization for variational image registration. IEEE Trans. Image Processing **22**, 4551–4559 (2013)
11. N. Chumchob, K. Chen, Improved variational image registration model and a fast algorithm for its numerical approximation. Numerical Methods for Partial Differential Equations **28**, 1966–1995 (2012)
12. N. Chumchob, K. Chen, C. Brito-Loeza, A fourth-order variational image registration model and its fast multigrid algorithm. Multiscale Modeling & Simulation **9**, 89–128 (2011)
13. M. Droske, W. Ring, A mumford-shah level-set approach for geometric image registration. SIAM journal on Applied Mathematics **66**, 2127–2148 (2006)
14. M. Droske, M. Rumpf, A variational approach to nonrigid morphological image registration. SIAM Journal on Applied Mathematics **64**, 668–687 (2004)
15. B. Fischer and J. Modersitzki, Fast diffusion registration, Contemp. Math., 313 (2002), pp. 117–129
16. B. Fischer, J. Modersitzki, Curvature based image registration. Journal of Mathematical Imaging and Vision **18**, 81–85 (2003)
17. B. Fischer, J. Modersitzki, Ill-posed medicine?an introduction to image registration. Inverse Probl. **24** (2008)
18. F. Gigengack, L. Ruthotto, M. Burger, C.H. Wolters, X. Jiang, K.P. Schafers, Motion correction in dual gated cardiac pet using mass-preserving image registration. IEEE transactions on medical imaging **31**, 698–712 (2012)
19. E. Haber, J. Modersitzki, Numerical methods for volume preserving image registration. Inverse problems **20**, 1621 (2004)
20. E. Haber, J. Modersitzki, Image registration with guaranteed displacement regularity. International Journal of Computer Vision **71**, 361–372 (2007)
21. S. Henn, A multigrid method for a fourth-order diffusion equation with application to image processing. SIAM Journal on Scientific Computing **27**, 831–849 (2005)
22. E. Hodneland, A. Lundervold, J. Rørvik, A.Z. Munthe-Kaas, Normalized gradient fields for nonlinear motion correction of dce-mri time series. Computerized Medical Imaging and Graphics **38**, 202–210 (2014)
23. W. Hu, Y. Xie, L. Li, and W. Zhang, A total variation based nonrigid image registration by combining parametric and non-parametric transformation models, Neurocomputing, 144 (2014), pp. 222–237
24. M. Ibrahim, K. Chen, C. Brito-Loeza, A novel variational model for image registration using gaussian curvature. Geometry, Imaging and Computing **1**, 417–446 (2014)
25. C.T. Kelley, *Iterative Methods for Optimization*, vol. 18 (Siam, 1999)
26. L. König, J. Rühaak, A fast and accurate parallel algorithm for non-linear image registration using normalized gradient fields, in Biomedical Imaging (ISBI), in 11th International Symposium on IEEE, vol. 2014 (IEEE, 2014), pp. 580–583

27. D. Loeckx, P. Slagmolen, F. Maes, D. Vandermeulen, P. Suetens, Nonrigid image registration using conditional mutual information. IEEE transactions on medical imaging **29**, 19–29 (2010)
28. L.M. Lui, T.C. Ng, A splitting method for diffeomorphism optimization problem using beltrami coefficients. Journal of Scientific Computing **63**, 573–611 (2015)
29. F. Maes, A. Collignon, D. Vandermeulen, G. Marchal, P. Suetens, Multimodality image registration by maximization of mutual information. IEEE transactions on Medical Imaging **16**, 187–198 (1997)
30. A. Mang, G. Biros, An inexact newton-krylov algorithm for constrained diffeomorphic image registration. SIAM journal on imaging sciences **8**, 1030–1069 (2015)
31. A. Mang, G. Biros, Constrained h1-regularization schemes for diffeomorphic image registration. SIAM journal on imaging sciences **9**, 1154–1194 (2016)
32. J. Modersitzki, *Numerical Methods for Image Registration* (Oxford University Press on Demand, 2004)
33. J. Modersitzki, *FAIR: Flexible Algorithms for Image Registration*, vol. 6 (SIAM, 2009)
34. J. Nocedal, S.J. Wright, *Numerical Optimization 2nd* (2006)
35. F.P. Oliveira, J.M.R. Tavares, Medical image registration: a review. Computer methods in biomechanics and biomedical engineering **17**, 73–93 (2014)
36. C.C. Paige, M.A. Saunders, Solution of sparse indefinite systems of linear equations. SIAM journal on numerical analysis **12**, 617–629 (1975)
37. J.P. Pluim, J.A. Maintz, M.A. Viergever, Mutual-information-based registration of medical images: a survey. IEEE transactions on medical imaging **22**, 986–1004 (2003)
38. G. Roland, L.T. Patrick, *Augmented Lagrangian and Operator-Splitting Methods in Nonlinear Mechanics* (SIAM, 1989)
39. J. Rühaak, L. König, M. Hallmann, N. Papenberg, S. Heldmann, H. Schumacher, B. Fischer, A fully parallel algorithm for multimodal image registration using normalized gradient fields, in Biomedical Imaging (ISBI), in *10th International Symposium on IEEE*, vol. 2013 (IEEE, 2013), pp. 572–575
40. A. Sotiras, C. Davatzikos, N. Paragios, Deformable medical image registration: A survey. IEEE transactions on medical imaging **32**, 1153–1190 (2013)
41. W. Sun, Y.-X. Yuan, *Optimization Theory and Methods: Nonlinear Programming*, vol. 1 (Springer Science & Business Media, 2006)
42. X.-C. Tai, J. Hahn, G.J. Chung, A fast algorithm for Euler's elastica model using augmented lagrangian method. SIAM Journal on Imaging Sciences **4**, 313–344 (2011)
43. A. Theljani, K. Chen, An augmented Lagrangian method for solving a new variational model based on gradients similarity measures and high order regulariation for multimodality registration. Inverse Problems and Imaging **13**, 309–335 (2019)
44. A. Theljani, K. Chen, A nash game based variational model for joint image intensity correction and registration to deal with varying illumination. Inverse Probl. **36** (2020)
45. A. Theljani, K. Chen, An unsupervised deep learning method for diffeomorphic mono-and multi-modal image registration, in *Medical Image Understanding and Analysis (MIUA 2019)*, vol. 1065 (Communications in Computer and Information Science, Springer, 2020)
46. P. Viola, W.M. Wells III, Alignment by maximization of mutual information. Int. J. Comput. Vis. **24**, 137–154 (1997)
47. C. Xing, P. Qiu, Intensity-based image registration by nonparametric local smoothing. IEEE Transactions on Pattern Analysis and Machine Intelligence **33**, 2081–2092 (2011)
48. D. Zhang, K. Chen, A novel diffeomorphic model for image registration and its algorithm. Journal of Mathematical Imaging and Vision **60**, 1261–1283 (2018)
49. D. Zhang, A. Theljani, K. Chen, On a new diffeomorphic multi-modality image registration model and its convergent gauss-newton solver. Journal of Mathematical Research with Applications **39**, 633–656 (2019)
50. J. Zhang, K. Chen, Variational image registration by a total fractional-order variation model. Journal of Computational Physics **293**, 442–461 (2015)
51. J. Zhang, K. Chen, B. Yu, An improved discontinuity-preserving image registration model and its fast algorithm. Applied Mathematical Modelling **40**, 10740–10759 (2016)

52. J. Zhang, K. Chen, and B. Yu, A novel high-order functional based image registration model with inequality constraint, Computers & Mathematics with Applications, 72 (2016), pp. 2887–2899

# Fast Algorithms for Surface Reconstruction from Point Cloud

**Yuchen He, Martin Huska, Sung Ha Kang, and Hao Liu**

**Abstract** We consider constructing a surface from a given set of point cloud data. We explore two fast algorithms to minimize the weighted minimum surface energy in [Zhao, Osher, Merriman and Kang, Comp Vision and Image Under, 80(3):295–319, 2000]. An approach using Semi-Implicit Method (SIM) improves the computational efficiency through relaxation on the time-step constraint. An approach based on Augmented Lagrangian Method (ALM) reduces the run-time via an Alternating Direction Method of Multipliers-type algorithm, where each sub-problem is solved efficiently. We analyze the effects of the parameters on the level-set evolution and explore the connection between these two approaches. We present numerical examples to validate our algorithms in terms of their accuracy and efficiency.

Keyword Surface reconstruction, Semi-implicit method, Augmented Lagrangian method, Point cloud

## 1 Introduction

Acquisition, creation and processing of 3D digital objects is an important topic in various fields, e.g., medical imaging [22], computer graphics [8, 11], industry [4], and preservation of cultural heritage [16]. A fundamental step is to reconstruct a

Y. He · S. H. Kang (✉) · H. Liu
School of Mathematics, Georgia Institute of Mathematics, Atlanta, USA
e-mail: kang@math.gatech.edu

Y. He
e-mail: royarthur@gatech.edu

H. Liu
e-mail: hao.liu@math.gatech.edu

M. Huska
Department of Mathematics, University of Bologna, Bologna, Italy
e-mail: martin.huska@unibo.it

 
X.-C. Tai et al. (eds.), *Mathematical Methods in Image Processing and Inverse Problems*, Springer Proceedings in Mathematics & Statistics 360,
https://doi.org/10.1007/978-981-16-2701-9_4

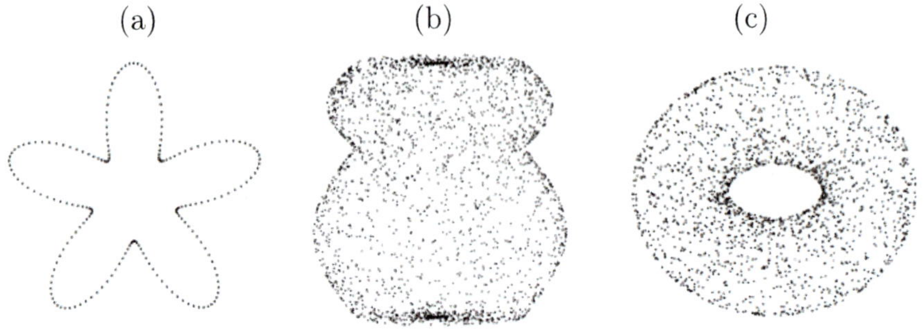


**Fig. 1** Test point clouds. **a** Five-fold circle (200 points). **b** Jar (2100 points). **c** Torus (2000 points)

surface from a set of point cloud data [1], denoted by $\mathcal{D} \subseteq \mathbb{R}^m$ for $m = 2$ or 3, such as in Fig. 1.

We focus on reconstructing a submanifold of codimension 1, denoted by $\Gamma$, i.e., a curve in $\mathbb{R}^2$ or a surface in $\mathbb{R}^3$, from the point cloud $\mathcal{D}$. We assume only the point locations are given, and no other geometrical information such as normal vectors at each point is known. We explore fast algorithms for minimizing the following energy proposed in [38]:

$$E_p(\Gamma) = \left( \int_\Gamma |d(\mathbf{x})|^p \, \mathrm{d}\mathbf{x} \right)^{\frac{1}{p}}, \tag{1}$$

where $d(\mathbf{x}) = \min_{\mathbf{y} \in \mathcal{D}} \{|\mathbf{x} - \mathbf{y}|\}$ is the distance from an arbitrary point $\mathbf{x} \in \mathbb{R}^m$ to $\mathcal{D}$, $p$ is a positive integer, and $\mathrm{d}\mathbf{x}$ is the surface area element. This energy is the $p$-norm of the distance function restricted on $\Gamma$. In [38], the authors used the fast sweeping scheme to compute the distances, and the associated Euler-Lagrange equations are solved by a gradient descent algorithm.

Among many ways to represent the underlying surface, e.g., (moving) least square projection [2, 30], radial basis function [9–11, 13, 18], poisson reconstruction [5, 14, 20, 21, 24], we use the level set method as in [15, 37, 38]. The level set formulation allows topological changes as well as self-intersection during the evolution [28, 29] and has gained popularity in many applications [12, 31, 36]. We represent the surface as a zero level set of $\phi : \mathbb{R}^m \to \mathbb{R}$:

$$\Gamma = \phi^{-1}(0) = \{\mathbf{x} \in \mathbb{R}^m \mid \phi(\mathbf{x}) = 0\}.$$

There are various related works on surface reconstruction from point cloud data: a convection model proposed in [37], a data-driven logarithmic prior for noisy data in [33], using surface tension to enrich the Euler-Lagrange equations in [17], and using principal component analysis to reconstruct curves embedded in sub-manifolds in [27]. A semi-implicit scheme is introduced in [34] to simulate the curvature and surface diffusion motion of the interface. In [25], the authors defined the surface via a collection of anisotropic Gaussians centered at each entry of the input point cloud, and used TVG-L1 model [7] for minimization. A similar strategy addresses an $\ell_0$ gradient regularization model in [23]. Some models incorporate additional information. In [26], the authors proposed a novel variational model, consisting of the distance, the normal, and the smoothness term. Euler's Elastica model is incorporated for surface reconstruction in [32] where graph cuts algorithm is used. The model in [15] extends the active contours segmentation model to 3D and implicitly allows to control the curvature of the level set function.

In this paper, we explore fast algorithms to minimize the weighted minimum surface energy (1) for $p = 1$ and 2. We propose a Semi-Implicit Method (SIM) to relax the time-step constraint for $p = 2$, and an Augmented Lagrangian Method (ALM) based on the alternating direction method of multipliers (ADMM) approach for $p = 1$. These algorithms minimize the weighted minimal surface energy (1) with high accuracy and superior efficiency. We analyze the behaviors of ALM in terms of the parameter choices and explore its connection to SIM. Various numerical experiments are presented to discuss the effects of the algorithms.

We organize this paper as follows. In Sect. 2, we present the two methods: SIM and ALM, and we explore their connection. Numerical experiments are presented in Sect. 3, and effects of the parameters are discussed in Sect. 3.3. We conclude our paper in Sect. 4.

## 2 Proposed Algorithms

Let $\Omega \subset \mathbb{R}^m$ ($m = 2$ or 3) denote a bounded domain containing the given point cloud data $\mathcal{D}$, a finite set of points. Using the level-set formulation for a codimension 1 submanifold $\Gamma$, the $d$-weighted minimum surface energy (1) can be rewritten as:

$$E_p(\phi) = \left( \int_\Omega |d(\mathbf{x})|^p \delta(\phi) |\nabla \phi| \, \mathbf{dx} \right)^{\frac{1}{p}}. \tag{2}$$

Here $\delta(x)$ is the Dirac delta function which takes $+\infty$ when $x = 0$, and 0 elsewhere. Compared to (1), this integral is defined on $\Omega$, which makes the computation flexible and free from explicitly tracking $\Gamma$. We use $p = 2$ for SIM introduced in Sect. 2.1, and $p = 1$ for ALM in Sect. 2.2. In general, $p = 2$ is a natural choice, since it provides better stability and efficiency for a semi-implicit type PDE-based method. For ALM, we explore $p = 1$ to take advantage of an aspect of fast algorithm in ADMM setting such as shrinkage, similarly to the case in [3]. Visually, the numerical results of surface reconstruction are similar for $p = 1$ or $p = 2$ (see Sect. 3).

## 2.1 Semi-implicit Method (SIM) to Minimize $\mathbf{E_2}$

We introduce a gradient-flow-based semi-implicit method to minimize

$$E_2(\phi) = \left(\int_\Omega d^2(\mathbf{x})\delta(\phi)|\nabla\phi|\,d\mathbf{x}\right)^{\frac{1}{2}}. \tag{3}$$

Following [38], the first variation of $E_2(\phi)$ with respect to $\phi$ is characterized as a functional:

$$\left\langle \frac{\partial E_2(\phi)}{\partial\phi}, v\right\rangle = -\int_\Omega \frac{1}{2}\delta(\phi)\left[\int_\Omega d^2(\mathbf{x})\delta(\phi)|\nabla\phi|\,d\mathbf{x}\right]^{-1/2} \nabla\cdot\left[d^2(\mathbf{x})\frac{\nabla\phi}{|\nabla\phi|}\right] v\,d\mathbf{x}$$
$$+\int_{\partial\Omega} \frac{d^2(\mathbf{x})\delta(\phi)}{|\nabla\phi|}(\nabla\phi\cdot\mathbf{n})v\,d\mathbf{x}$$

for any test function $v$ from the Sobolev space $H^1$ where $\mathbf{n}$ denotes the outward normal direction along $\partial\Omega$. Minimizing (3) is equivalent to finding the critical point $\phi$ such that $\left\langle \frac{\partial E_2(\phi)}{\partial\phi}, v\right\rangle = 0, \forall v \in H^1$. This is associated with solving the following initial value problem:

$$\begin{cases} \dfrac{\partial\phi}{\partial t} = \bar{f}(d,\phi)\nabla\cdot\left[d^2(\mathbf{x})\dfrac{\nabla\phi}{|\nabla\phi|}\right] \text{ in } \Omega, \\ \phi(\mathbf{x},0) = \phi^0, \end{cases} \tag{4}$$

where $\phi^0$ is an initial guess for the unknown $\phi$, and $\bar{f}(d,\phi) = \frac{1}{2}\delta(\phi)\left[\int_\Omega d^2(\mathbf{x})\delta(\phi)|\nabla\phi|\,d\mathbf{x}\right]^{-1/2}$ with a boundary condition $\frac{d^2(\mathbf{x})\delta(\phi)}{|\nabla\phi|}\frac{\partial\phi}{\partial\mathbf{n}} = 0$ on $\partial\Omega$. The steady state solution of (4) gives a minimizer $\phi^*$ of $E_2(\phi)$. Since our focus is on the zero level set of $\phi$, we apply reinitialization to $\phi$ after every several iterations to make our scheme more stable. This modifies $\phi$ to be a signed distance function while keeping the location of the zero level set (see Sect. 3.1). Thus, the effect of the boundary condition is negligible away from the boundary of the image domain. To utilize the Fast Fourier Transform, we apply periodic boundary condition for computation.

Here the delta function $\delta$ is realized as the derivative of the one dimensional Heaviside function $H : \mathbb{R} \to \{0, 1\}$. We adopt the smooth approximation of $H(\phi)$ as in [6]:

$$H(\phi) \approx H_\varepsilon(\phi) = \frac{1}{2} + \arctan(\phi/\varepsilon)/\pi \quad\text{and}\quad \delta(\phi) \approx H'_\varepsilon(\phi) = \frac{\varepsilon}{\pi(\varepsilon^2+\phi^2)} \tag{5}$$

with $\varepsilon > 0$ as the smoothness parameter. Then $\bar{f}$ is approximated by its smoothed version $f$ expressed as

$$f(d,\phi) = \frac{1}{2}\frac{\varepsilon}{\pi(\varepsilon^2+\phi^2)}\left[\int_\Omega d^2(\mathbf{x})\frac{\varepsilon}{\pi(\varepsilon^2+\phi^2)}|\nabla\phi|\,\mathbf{dx}\right]^{-1/2}.$$

We add a stabilizing diffusive term $-\beta\Delta\phi$ for $\beta > 0$ on both sides of the PDE in (4) to consolidate the computation, similarly to [34]:

$$\frac{\partial\phi}{\partial t} - \beta\Delta\phi = -\beta\Delta\phi + f(d,\phi)\nabla\cdot\left[d^2(\mathbf{x})\frac{\nabla\phi}{|\nabla\phi|}\right]. \tag{6}$$

Employing a semi-implicit scheme, we solve $\phi$ from (6) by iteratively updating $\phi^{n+1}$ using $\phi^n$ via the following equation:

$$\frac{\phi^{n+1}}{\Delta t} - \beta\Delta\phi^{n+1} = \frac{\phi^n}{\Delta t} - \beta\Delta\phi^n + f(d,\phi^n)\nabla\cdot\left[d^2(\mathbf{x})\frac{\nabla\phi^n}{|\nabla\phi^n|}\right], \tag{7}$$

where $\Delta t$ is the time-step. This equation can be efficiently solved by the Fast Fourier Transform (FFT). Denoting the discrete Fourier transform by $\mathcal{F}$ and its inverse by $\mathcal{F}^{-1}$, we have

$$\begin{aligned}\mathcal{F}(\phi)(i\pm1,j) &= e^{\pm 2\pi\sqrt{-1}(i-1)/M}\mathcal{F}(\phi)(i,j),\\ \mathcal{F}(\phi)(i,j\pm1) &= e^{\pm 2\pi\sqrt{-1}(j-1)/N}\mathcal{F}(\phi)(i,j).\end{aligned}$$

Accordingly, the discrete Fourier transform of $\Delta\phi$ is

$$\mathcal{F}(\Delta\phi)(i,j) = \left[2\cos(\pi\sqrt{-1}(i-1)/M) + 2\cos(\pi\sqrt{-1}(j-1)/N) - 4\right]\mathcal{F}\phi(i,j).$$

Here the coefficient in front of $\mathcal{F}\phi(i,j)$ represents the diagonalized discrete Laplacian operator in the frequency domain. Let $g_1$ be the right side of (7), then the solution $\phi^{n+1}(i,j)$ of (7) is computed via

$$\phi^{n+1}(i,j) = \mathcal{F}^{-1}\left(\frac{\mathcal{F}(g_1)(i,j)}{(1-\beta\Delta t\left[2\cos(\pi\sqrt{-1}(i-1)/M) + 2\cos(\pi\sqrt{-1}(j-1)/N) - 4\right])}\right). \tag{8}$$

As for the stopping criterion, we exploit the mean relative change of the weighted minimum surface energy (1). At the $n$th iteration, the algorithm terminates if

$$\frac{|\bar{e}_{n-1}^k - \bar{e}_n^k|}{\bar{e}_n^k} < 10^{-4}, \quad \text{where} \quad \bar{e}_n^k = \frac{1}{k}\sum_{i=n-k}^{n} E_p(\phi^i). \tag{9}$$

Here the quantity $\bar{e}_n^k$ represents the average of the energy values computed from the $(n-k)$th to the $n$th iteration for some $k \in \mathbb{N}, k \geq 1$. We fix $k = 10$ and set $p = 2$ for SIM. We summarize the main steps of SIM in Algorithm 1.

---

**Algorithm 1:** SIM for the weighted minimum surface (3)

---

**Initialization**: $d$, $\phi^0$ and $n = 0$.
**while** *the stopping criterion* (9) *with* $p = 2$ *is greater than* $10^{-4}$ **do**
  Update $\phi^{n+1}$ from $\phi^n$ solving (8);
  Update $n \leftarrow n + 1$;
**end**
**Output**: $\phi^n$ such that $\{\phi^n = 0\}$ approximates $\{\phi^* = 0\}$.

---

## 2.2 *Augmented Lagrangian Method (ALM) to Minimize* $\mathbf{E}_1$

In this section, we present an augmented Lagrangian-based method to minimize the weighted minimum surface energy (2) for $p = 1$, i.e.,

$$E_1(\phi) = \int_\Omega d(\mathbf{x})\delta(\phi)|\nabla\phi| \, \mathrm{d}\mathbf{x}. \tag{10}$$

For the non-differentiable term $|\nabla\phi|$ in (10), we utilize the variable-splitting technique and introduce an auxiliary variable $\mathbf{p} = \nabla\phi$. We rephrase the minimization of $E_1(\phi)$ as a constrained optimization problem:

$$\{\phi^*, \mathbf{p}^*\} = \arg\min_{\phi,\mathbf{p}} \int_\Omega \frac{\varepsilon d|\mathbf{p}|}{\pi(\varepsilon^2 + \phi^2)} \mathrm{d}\mathbf{x}, \quad \text{subject to } \mathbf{p} = \nabla\phi, \tag{11}$$

here we replace $\delta(\phi)$ by its smooth approximation $H'_\varepsilon(\phi)$ as in (5). To solve problem (11), we formulate the augmented Lagrangian functional:

$$\mathcal{L}(\phi, \mathbf{p}, \boldsymbol{\lambda}; r) = \int_\Omega \frac{\varepsilon d|\mathbf{p}|}{\pi(\varepsilon^2 + \phi^2)} \mathrm{d}\mathbf{x} + \frac{r}{2}\int_\Omega |\mathbf{p} - \nabla\phi|^2 \, \mathrm{d}\mathbf{x} + \int_\Omega \boldsymbol{\lambda} \cdot (\mathbf{p} - \nabla\phi) \, \mathrm{d}\mathbf{x}, \tag{12}$$

where $r > 0$ is a scalar penalty parameter and $\boldsymbol{\lambda} : \mathbb{R}^m \to \mathbb{R}^m$ represents the Lagrangian multiplier. Minimizing (12) amounts to considering the following saddle-point problem:

$$
\begin{aligned}
&\text{Find } (\phi^*, \mathbf{p}^*, \boldsymbol{\lambda}^*) \in \mathbb{R} \times \mathbb{R}^m \times \mathbb{R}^m \\
&\quad \text{s.t. } \mathcal{L}(\phi^*, \mathbf{p}^*, \boldsymbol{\lambda}; r) \leq \mathcal{L}(\phi^*, \mathbf{p}^*, \boldsymbol{\lambda}^*; r) \leq \mathcal{L}(\phi, \mathbf{p}, \boldsymbol{\lambda}^*; r); \\
&\quad \forall (\phi, \mathbf{p}, \boldsymbol{\lambda}) \in \mathbb{R} \times \mathbb{R}^m \times \mathbb{R}^m.
\end{aligned} \tag{13}
$$

Given $\phi^n$, $\mathbf{p}^n$, and $\boldsymbol{\lambda}^n$, for $n = 0, 1, 2, \ldots$, the $(n+1)$th iteration of an ADMM-type algorithm for (13) consists of solving a series of sub-problems:

$$
\phi^{n+1} = \arg\min_{\phi} \mathcal{L}(\phi, \mathbf{p}^n, \boldsymbol{\lambda}^n; r); \tag{14}
$$

$$
\mathbf{p}^{n+1} = \arg\min_{\mathbf{p}} \mathcal{L}(\phi^{n+1}, \mathbf{p}, \boldsymbol{\lambda}^n; r); \tag{15}
$$

$$
\boldsymbol{\lambda}^{n+1} = \boldsymbol{\lambda}^n + r\left(\mathbf{p}^{n+1} - \nabla \phi^{n+1}\right). \tag{16}
$$

Each sub-problem can be solved efficiently. First, we find the minimizer of the sub-problem (14) by solving its Euler-Lagrange equation:

$$
-r\Delta\phi^{n+1} = \frac{2d\varepsilon|\mathbf{p}^n|\phi^n}{\pi(\varepsilon^2 + (\phi^n)^2)^2} - \nabla \cdot (r\mathbf{p}^n + \boldsymbol{\lambda}^n). \tag{17}
$$

Here $\Delta$ is the Laplacian operator. Following [3], we introduce a frozen-coefficient term $\eta\phi$, for $\eta > 0$, on both sides of (17) to stabilize the computation; thus, (14) is solved using the following equation:

$$
\eta\phi^{n+1} - r\Delta\phi^{n+1} = \eta\phi^n + \frac{2d\varepsilon|\mathbf{p}^n|\phi^n}{\pi(\varepsilon^2 + (\phi^n)^2)^2} - \nabla \cdot (r\mathbf{p}^n + \boldsymbol{\lambda}^n). \tag{18}
$$

We solve this via FFT, similarly to (8) for SIM. Thus, the $\phi$ sub-problem is solved via

$$
\begin{aligned}
&\phi^{n+1}(i, j) \\
&= \mathcal{F}^{-1}\left(\frac{\mathcal{F}(g_2)(i, j)}{\left(\eta - r\left[2\cos(\pi\sqrt{-1}(i-1)/M) + 2\cos(\pi\sqrt{-1}(j-1)/N) - 4\right]\right)}\right).
\end{aligned} \tag{19}
$$

Second, the $\mathbf{p}$ sub-problem (15) is equivalent to a weighted Total Variation (TV) minimization, whose solution admits a closed-form expression using the shrinkage operator [35]. Explicitly, the updated $\mathbf{p}^{n+1}$ is computed via:

$$
\mathbf{p}^{n+1} = \max\left\{0, 1 - \frac{d\,\varepsilon}{\pi(\varepsilon^2 + (\phi^{n+1})^2)|r\nabla\phi^{n+1} - \boldsymbol{\lambda}^n|}\right\}\left(\nabla\phi^{n+1} - \frac{\boldsymbol{\lambda}^n}{r}\right). \tag{20}
$$

Finally, the Lagrangian multiplier $\boldsymbol{\lambda}$ is updated by (16). The stopping criterion for the ALM iteration is the same as that for SIM (9), but with $p = 1$. We summarize the main steps of ALM in Algorithm 2.

**Algorithm 2:** ALM for the weighted minimum surface (10)

**Initialization**: $d, r, \phi^0, \mathbf{p}^0, \boldsymbol{\lambda}^0$, and $n = 0$.
**while** *the stopping criterion* (9) *with* $p = 1$ *is greater than* $10^{-4}$ **do**
  Update $\phi^{n+1} = \arg\min_\phi \mathcal{L}(\phi, \mathbf{p}^n, \boldsymbol{\lambda}^n; r)$ via (19) ;
  Update $\mathbf{p}^{n+1} = \arg\min_{\mathbf{p}} \mathcal{L}(\phi^{n+1}, \mathbf{p}, \boldsymbol{\lambda}^n; r)$ via (20);
  Update $\boldsymbol{\lambda}^{n+1} = \boldsymbol{\lambda}^n + r(\mathbf{p}^{n+1} - \nabla\phi^{n+1})$;
  Update $n \leftarrow n + 1$;
**end**
**Output**: $\phi^n$ such that $\{\phi^n = 0\}$ approximates $\{\phi^* = 0\}$.

## 2.3 Connection Between SIM and ALM Algorithms

Note that both SIM and ALM involve solving elliptic PDEs of the form:

$$a\phi - b\Delta\phi = g, \tag{21}$$

for some constants $a, b > 0$, and a function $g$ defined on $\Omega$. For SIM, it is equation (7):

$$\underbrace{\frac{1}{\Delta t}}_{a}\phi^{n+1} - \underbrace{\beta}_{b}\Delta\phi^{n+1} = \underbrace{\frac{\phi^n}{\Delta t} - \beta\Delta\phi^n + f(d, \phi^n)\nabla \cdot \left[d^2(\mathbf{x})\frac{\nabla\phi^n}{|\nabla\phi^n|}\right]}_{g},$$

and for ALM, it is equation (18):

$$\underbrace{\eta}_{a}\phi^{n+1} - \underbrace{r}_{b}\Delta\phi^{n+1} = \underbrace{\eta\phi^n + \frac{2d\varepsilon|\mathbf{p}^n|\phi^n}{\pi(\varepsilon^2 + (\phi^n)^2)^2} - \nabla \cdot (r\mathbf{p}^n + \boldsymbol{\lambda}^n)}_{g}.$$

We remark interesting connections between SIM and ALM. First, both methods have stabilizing terms but in different positions on the left side of (21). For SIM, it is $-\beta\Delta\phi$, while for ALM, it is $\eta\phi$. Second, relating the coefficients of $\phi$, $1/\Delta t$ in SIM gives insight to the effect of $\eta$ in ALM. In general, a large $\eta$ slows down the convergence of ALM, while a small $\eta$ accelerates it (as the effect of $\frac{1}{\Delta t}$ on SIM). Figure 2 shows convergence behaviors of ALM for different $\eta$, using the five-fold circle point cloud in Fig. 1a. It displays the CPU time (in seconds) for $r = 1$, $\varepsilon = 1$, and $\eta$ varying from 0.05 to 0.5. Note that as $\eta$ increases, the time required to reach the convergence increases almost quadratically at first, then stays around the same level. Third, the correspondence between $b = \beta$ in SIM, and $b = r$ in ALM allows another interpretation of the parameter $r$ in ALM. In SIM, a large $\beta$ smears the solution and avoids discontinuities or sharp corners, and for ALM, a large $r$ also allows to pass through fine details. Figure 7 in Sect. 3 presents more details, where we experiment with different $r$ and $\varepsilon$ values for the five-fold circle point cloud shown in Fig. 1a.

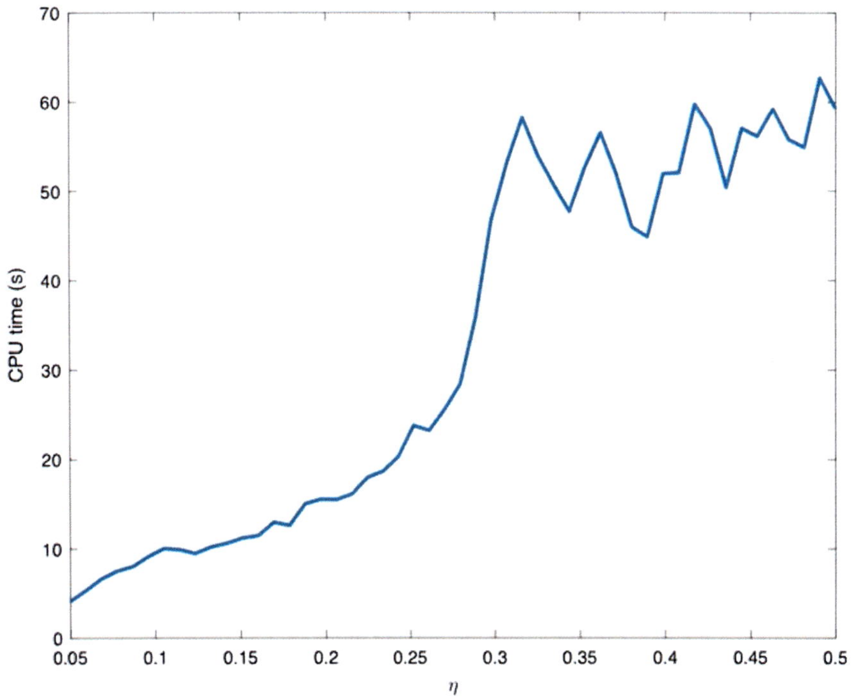


**Fig. 2** The CPU-time (s) of ALM until convergence for the five-fold circle point cloud in Fig. 1a. Here $r = \varepsilon = 1$ and $\eta$ varies from 0.05 to 0.5. The connection between SIM and ALM indicates that a large $\eta$ slows down ALM. In this graph, as $\eta$ increases, the time required to reach the convergence increases

# 3 Numerical Implementations, Experiments and Effects of Parameters

In this section, we describe the implementation details and present numerical experiments. For both SIM and ALM, we vary the value of $\varepsilon$ from 0.5 to 1. For SIM, we use $\Delta t = 500$. When the point cloud $\mathcal{D}$ is in 2D, we set $\beta = 0.1$, and when $\mathcal{D}$ is in 3D, $\beta = 0.01$. For ALM, the value of $\eta$ ranges from 0.05 to 1, and $r$ from 0.5 to 2.

The code is written in MATLAB and executed without additional machine support, e.g. parallelization nor GPU-enhanced computations. All the experiments are performed on Intel® Core™4-Core 1.8 GHz (4.0 GHz with Turbo) machine, with 16 GB/RAM and Intel® UHD Graphics 620 graphic card under Windows OS. The contours and isosurfaces are displayed using MATLAB visualization engine. No post-processing, e.g., smoothing nor sharpening, is applied.

### 3.1 Implementation Details

We illustrate the details for planar point clouds, i.e., $\mathcal{D} \subseteq \mathbb{R}^2$, while the extension to $\mathbb{R}^3$ is straightforward. Let the computational domain $\Omega = [0, M] \times [0, N]$, $M, N > 0$, be discretized by a Cartesian grid with $\Delta x = \Delta y = 1$. When the input point cloud $\mathcal{D}$ requires different values for $\Delta x$ and $\Delta y$, one can consider the density features of $\mathcal{D}$, e.g., using the local sample density defined as the radius of the largest inner empty disk: $h_{\mathcal{D}} = \sup_{\mathbf{x}\in\Omega} \min_{1<i<|\mathcal{D}|} \|\mathbf{x} - y_i\|_2$, one can scale the data $\mathcal{D}$ up (or down) such that $h_{\mathcal{D}} \geq 0.5$ which allows the discretization step $\Delta x = \Delta y = 1$ being sufficient in capturing a water-tight surface. After the computation, we transform the reconstructed surface back to the original scale. For any function $u$ (or a vector field $\mathbf{v} = (v^1, v^2)$) defined on $\Omega$, we use $u_{i,j}$ or $u(i, j)$ to denote $u(i\Delta x, i\Delta y)$. We use the usual backward and forward finite difference schemes:

$$\partial_1^- u_{i,j} = \begin{cases} u_{i,j} - u_{i-1,j}, & 1 < i \leq M; \\ u_{1,j} - u_{M,j}, & i = 1. \end{cases} \qquad \partial_1^+ u_{i,j} = \begin{cases} u_{i+1,j} - u_{i,j}, & 1 \leq i < M - 1; \\ u_{1,j} - u_{M,j}, & i = M. \end{cases}$$

$$\partial_2^- u_{i,j} = \begin{cases} u_{i,j} - u_{i,j-1}, & 1 < j \leq N; \\ u_{i,1} - u_{i,N}, & j = 1. \end{cases} \qquad \partial_2^+ u_{i,j} = \begin{cases} u_{i,j+1} - u_{i,j}, & 1 \leq j < N - 1; \\ u_{i,1} - u_{i,N}, & j = N. \end{cases}$$

The gradient, divergence and the Laplacian operators are approximated as follows:

$$\begin{aligned} \nabla u_{i,j} &= ((\partial_1^- u_{i,j} + \partial_1^+ u_{i,j})/2, (\partial_2^- u_{i,j} + \partial_2^+ u_{i,j})/2); \\ \nabla \cdot \mathbf{v}_{i,j} &= (\partial_1^+ v_{i,j}^1 + \partial_1^- v_{i,j}^1)/2 + (\partial_2^+ v_{i,j}^2 + \partial_2^- v_{i,j}^2)/2; \\ \Delta u_{i,j} &= \partial_1^+ u_{i,j} - \partial_1^- u_{i,j} + \partial_2^+ u_{i,j} - \partial_2^- u_{i,j}. \end{aligned}$$

The distance function $d$ is computed once at the beginning and no update is needed. It satisfies the Eikonal equation:

$$\begin{cases} |\nabla d| = 1 \text{ in } \Omega, \\ d(\mathbf{x}) = 0 \text{ for } \mathbf{x} \in \mathcal{D}, \end{cases} \tag{22}$$

and discretizing (22) via the Lax-Friedrich scheme leads to an updating formula:

$$d_{i,j}^{n+1} = \frac{1}{2}\left(1 - |\nabla d_{i,j}^n| + \frac{d_{i+1,j}^n + d_{i-1,j}^n}{2} + \frac{d_{i,j+1}^n + d_{i,j-1}^n}{2}\right). \tag{23}$$

We solve (23) using the fast sweeping method [19] with complexity $O(G)$ for $G$ grid points.

Keeping $\phi^n$ to be a signed distance function during the iteration improves the stability of level-set-based algorithms. We reinitialize $\phi^n$ at the $n$th iteration by solving the following PDE:

$$\begin{cases} \phi_\tau + \text{sign}(\phi)(|\nabla\phi| - 1) = 0, \\ \phi(\mathbf{x}, 0) = \phi^n. \end{cases} \tag{24}$$

Here the subscript $\tau$ represents the partial derivative with respect to an artificial time, and sign : $\mathbb{R} \to \{-1, 0, 1\}$ is the sign function. We discretize (24) via an explicit time Lax-Friedrichs scheme. For $k = 0, 1, \ldots, K$, we update

$$\begin{aligned} \phi_{i,j}^{(k+1)} &= \phi_{i,j}^{(k)} \\ &- \Delta\tau \left[ \text{sign}(\phi_{i,j}^{(k)})(|\nabla\phi_{i,j}^{(k)}| - 1) - \frac{\phi_{i-1,j}^k + \phi_{i+1,j}^k + \phi_{i,j-1}^k + \phi_{i,j+1}^k - 4\phi_{i,j}^k}{2} \right], \end{aligned} \tag{25}$$

with $\phi_{i,j}^{(0)} = \phi_{i,j}^n$. In practice, $\phi^n$ being a signed distance function near the 0-level-set is important; thus, it is sufficient to evolve (25) for a small $K$ and update $\phi^n$ with $\phi^{(K)}$. We fix $K = 10$ throughout this paper.

### 3.2 *Numerical Experiments of 2D and 3D Point Clouds*

For our first experiment, Fig. 3 displays a set of planar curves reconstructed from 2D point clouds confined within a square $\Omega = [0, 100]^2 \subset \mathbb{R}^2$. We generate the data using four different shapes: a triangle, an ellipse, a square whose corners are missing, and a five-fold circle. For these cases, we use a centered circle with radius 30 as the initial guess, shown in Fig. 3a. Figure 3b and c display the given $\mathcal{D}$, as well as the curves identified by SIM and ALM with $r = 1.5$, respectively. Both methods produce comparably accurate results. In the triangle example, corners get as close as the approximated delta function (with parameter $\varepsilon$) allows for both methods. The ellipse and square results fit very closely to the respective point clouds. For the five-fold-circle, there is a slight difference in how the curve fits the edges, yet the results are very compatible.

Table 1 shows the CPU times (in seconds) for SIM, ALM using $r = 0.5, 1, 1.5$, and 2, as well as the times for the explicit method in [38] using $\Delta t = 20$ on the same data sets. With proper choices of $r$, ALM outperforms the other methods in terms of computational efficiency. SIM is stable without any dependency on the choice of parameters, and its run-times are comparable to the best performances of ALM in most cases. Both methods are faster than the explicit method in all the examples.

The second set of experiments reconstruct surfaces from the point clouds in 3D: a jar in Fig. 1b and a torus in Fig. 1c within $\Omega = [0, 50]^3$. In Fig. 4, we show the reconstructed surfaces using SIM and ALM. A portion of the given point cloud is superposed for validation in each case. Both methods successfully capture the overall shapes and non-convex features of the jar, as well as the torus. There are only slight differences in the reconstruction between using SIM with $p = 2$ and using ALM with $p = 1$.

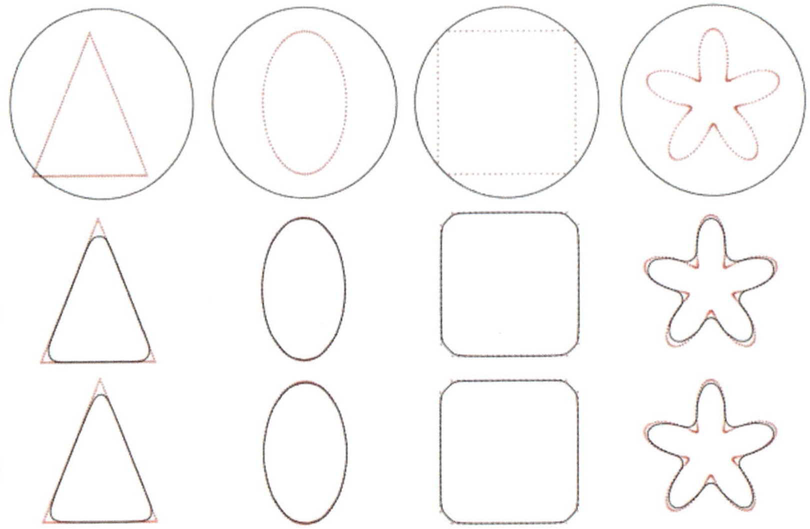


**Fig. 3** The test point clouds: triangle with 150 points, ellipse with 100 points, square with 80 points, and five-fold-circle with 200 points. **a** The top row, an identical initial condition applied to SIM and ALM for different $\mathcal{D}$. **b** The middle row, the results obtained by SIM. **c** The bottom row, the results obtained by ALM using $r = 1.5$. Both methods give compatible results

**Table 1** CPU time (s) for SIM, ALM using $r = 0.5$, 1, 1.5, and 2, and the explicit method in [38] with $\Delta t = 20$ for the point cloud data sets in Fig. 3. Both SIM and ALM shows fast convergence

| Object | ALM ($r = 0.5$) | ALM ($r = 1$) | ALM ($r = 1.5$) | ALM ($r = 2$) | SIM | [38] |
|---|---|---|---|---|---|---|
| Triangle | – | 1.45 | 1.31 | 1.48 | 1.50 | 5.25 |
| Ellipse | 1.22 | 1.03 | 1.33 | 1.37 | 1.49 | 3.89 |
| Square | – | – | 0.94 | 1.20 | 1.09 | 2.07 |
| Five-fold circle | 0.83 | 1.44 | 1.86 | 1.22 | 1.96 | 4.18 |

Table 2 shows the efficiency of SIM and ALM compared to the explicit method in [38] for the experiments in Fig. 4. Thanks to the semi-implicit scheme, the time step can be large and we used $\Delta t = 500$ in SIM; in the explicit method, we are forced to use much smaller time step $\Delta t = 20$ to maintain the stability. The improvement of run-time in ALM is carefully controlled by the parameters $r$, $\varepsilon$ and $\eta$. We choose $r = 1.3$, $\varepsilon = 0.5$ and $\eta = 0.6$ for both cases. Both SIM and ALM efficiently provide accurate reconstructions.

The third set of examples show the effect of the distance function $d$. Notice that the weighted minimal surface energy (1) is mainly driven by the distance function $d$, that is, the given point cloud $\mathcal{D}$ determines the landscape of $d$, which affects the behavior of the level-set during the evolution. Figure 5 shows the evolution using

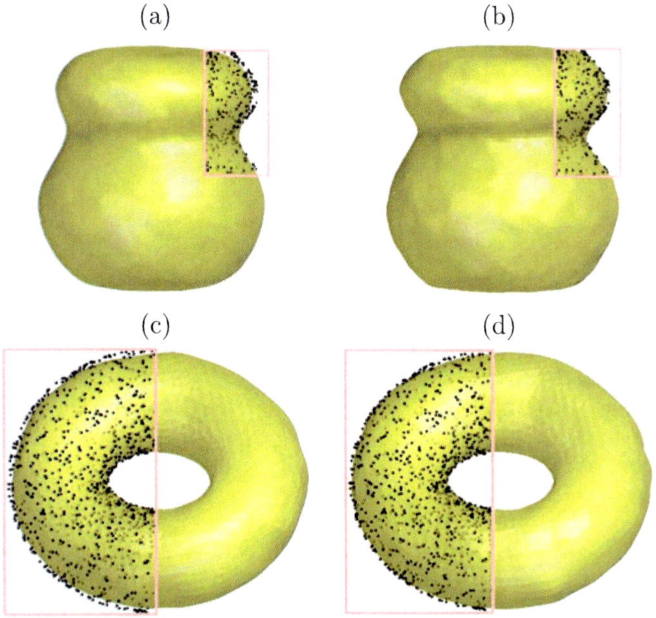


**Fig. 4** The first row shows ALM and SIM applied to the 3D jar point cloud in Fig. 1b. **a** The result of ALM with $r = 1.3$, $\varepsilon = 0.5$, $\eta = 0.6$. **b** The result of SIM. The second row shows the methods applied to the 3D torus point cloud in Fig. 1c. **c** The result of ALM with $r = 1.3$, $\varepsilon = 0.5$, $\eta = 0.6$. **d** The result of SIM. Both methods are compatible and show good results

**Table 2** CPU time (s) of SIM and ALM compared to the explicit method in [38] for the point cloud data sets of Fig. 4. Both SIM and ALM show fast convergence

| Object | ALM | SIM | [38] |
|---|---|---|---|
| Jar | 29.69 | 29.42 | 74.44 |
| Torus | 47.32 | 33.58 | 114.20 |

ALM, applied to different subsets of point clouds sampled from the same bunny face shape. The densities of the point cloud vary for the three different regions: the face with $n_1$ points, the head with $n_2$ points, and each ear with $n_3$ points. Figure 5 (a) shows the given point cloud for $(n_1, n_2, n_3) = (20, 10, 20)$, with the 0-level-set of $\phi^n$ at 15th iteration, (b) for $(n_1, n_2, n_3) = (50, 10, 20)$, at 18th iteration, and (c) for $(n_1, n_2, n_3) = (20, 10, 40)$, at 20th iteration. These three curves eventually degenerate to a point, since the energy model (2) drives curves to have short lengths, i.e., the level set tends to shrink. (d) for $(n_1, n_2, n_3) = (50, 10, 40)$ shows the converged solution. In (a)–(c), denser parts of the point cloud attract the curve with stronger forces, and the sparser parts of the point cloud fail to lock the curve. In (d), with a more balanced distribution of points, the curve converges to the correct shape.

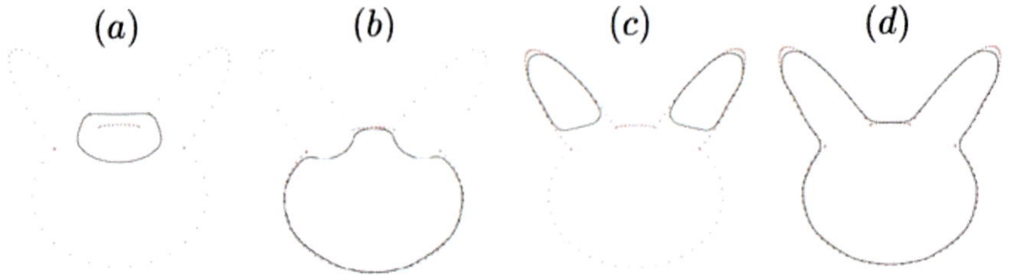


**Fig. 5** The effect of the distance function for varying-density point clouds: the face with $n_1$ points, the head with $n_2$ points, and each ear with $n_3$ points. **a** the given point cloud is with $(n_1, n_2, n_3) = (20, 10, 20)$, and shows the 0-level-set of $\phi^n$ at 15th iteration, **b** $(n_1, n_2, n_3) = (50, 10, 20)$, and shows 18th iteration, and **c** $(n_1, n_2, n_3) = (20, 10, 40)$, and shows 20th iteration. These three curves eventually degenerate to a point. **d** $(n_1, n_2, n_3) = (50, 10, 40)$ and shows the converged solution. The potential energy (1) is mainly driven by the distance function $d$, which affects the level-set evolution

The fourth set of examples demonstrate the robustness of ALM and SIM against noise. Figure 6 shows the reconstructed curves from clean and noisy data: (a)–(c) are results of ALM, and (d)–(f) are results of SIM. (a) and (d) in the first column show results obtained from the clean data, which has 200 points sampled from a three-fold circle. Gaussian noise with standard deviation 1 is added to both $x$ and $y$ coordinates to generate noisy point cloud in the second column, (b) and (e). To show the differences, the third column superposes both results reconstructed from clean and noisy point clouds. Both ALM and SIM provide compatible results. For the noisy data, although the reconstructed curves show some oscillations, they are very close to the solutions using the clean data, respectively.

### 3.3 *Choice of Parameters for ALM and the Effects*

The proposed ALM has one parameter $r > 0$, and the model (2) uses the delta function, where the smoothness parameter $\varepsilon > 0$ is added to stabilize the computation. Both parameters have straightforward effects on the level-set evolution from (17). For example, consider a set of points within a thin-band around the 0-level-set of $\phi^n$, denoted by $B_\varepsilon = \{\mathbf{x} \mid -2\varepsilon/\sqrt{3} < \phi^n(\mathbf{x}) < 2\varepsilon/\sqrt{3}\}$. By the continuity of $\phi^n$, there exist $\mathbf{y}$ and $\mathbf{z} \in B_\varepsilon$ such that $\phi^n(\mathbf{y}) = -\varepsilon/\sqrt{3}$ and $\phi^n(\mathbf{z}) = \varepsilon/\sqrt{3}$; these values are the minimum and maximum of the function $h(x) = \frac{2\varepsilon x}{\pi(\varepsilon^2+x^2)^2}$, respectively. At these points, (17) takes the following forms:

$$\Delta\phi^{n+1} = \begin{cases} -9\ d\ |\mathbf{p}^n|/(8\sqrt{3}\pi\ \varepsilon^2 r) + \nabla\cdot(\mathbf{p}^n + \frac{\lambda^n}{r}) & \text{at } \mathbf{y}. \\ 9\ d\ |\mathbf{p}^n|/(8\sqrt{3}\pi\ \varepsilon^2 r) + \nabla\cdot(\mathbf{p}^n + \frac{\lambda^n}{r}) & \text{at } \mathbf{z}. \end{cases} \tag{26}$$

The first terms in the right hand side of (26) show that with a smaller value of $\varepsilon$, there are less number of points in $B_\varepsilon$, but the influence from $d$ becomes stronger.

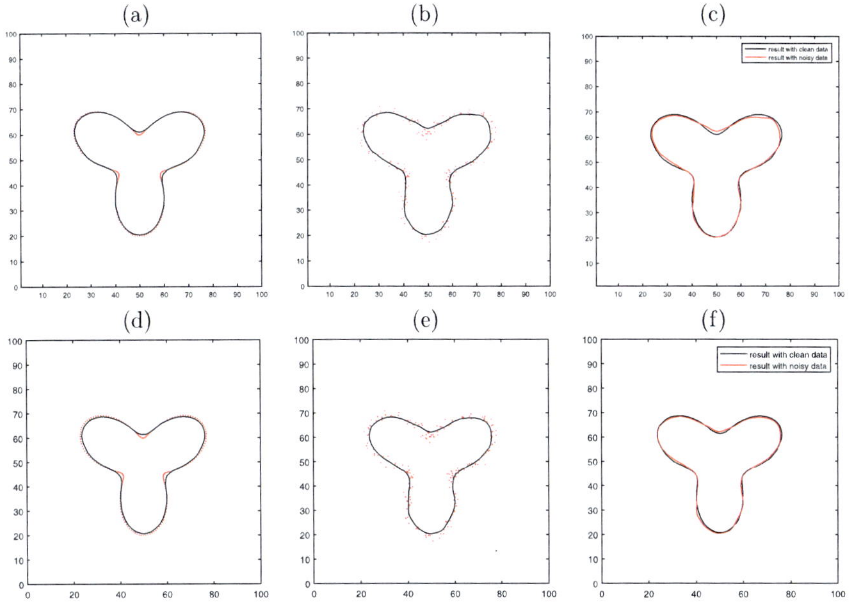


**Fig. 6** The influence of noise on reconstructing three-fold circle with 200 points: **a–c** ALM and **d–f** SIM. The first column shows the reconstructed curves from clean data, and the second column the reconstructions from noisy data. The third column shows the comparison between the two reconstructed curves in first two columns

With a larger value of $\varepsilon$, $d$ affects more number of points in $B_\varepsilon$, but with a weaker influence. Varying values of $r$ also modifies the effect of $d$, while the size of $B_\varepsilon$ is not changed.

We also find that $\varepsilon$ interacts with $r$ and effectively modifies the shape of the level-set. Figure 7 shows the results for ALM using different combinations of $r$ and $\varepsilon$, on the five-fold circle point cloud in Fig. 1a. For a fixed $r$, increasing $\varepsilon$ makes the approximated delta function smoother; consequently, narrow and elongated shapes are omitted, and the reconstructed curve becomes more convex. For a fixed $\varepsilon$, a larger $r$ causes loss of more details, as discussed in Sect. 2.3. The speed of convergence varies for different combinations of $r$ and $\varepsilon$. When the choices are reasonable, the algorithm converges fast within 2 s. When both $r$ and $\varepsilon$ are large, results are not as good, and the convergence is slow.

Another observation comes from (20). For any point $\mathbf{x}$ and $n \geq 0$, if the value

$$Q^n(\varepsilon, r) := \phi^n \pi |r \nabla \phi^n - \boldsymbol{\lambda}^{n-1}|\varepsilon^2 - d\varepsilon + (\phi^n)^3 \pi |r \nabla \phi^n - \boldsymbol{\lambda}^{n-1}|$$

is positive, then $\mathbf{p}^n(\mathbf{x}) = 0$, and $d$ has no direct effect on (18) at $\mathbf{x}$ in the next iteration. Regarding $Q^n(\varepsilon, r)$ as a quadratic polynomial in terms of $\varepsilon$ parameterized by $r$, the sign of $Q^n(\varepsilon, r)$ depends on the sign of $\phi^n(\mathbf{x})$ and the sign of its discriminant computed via:

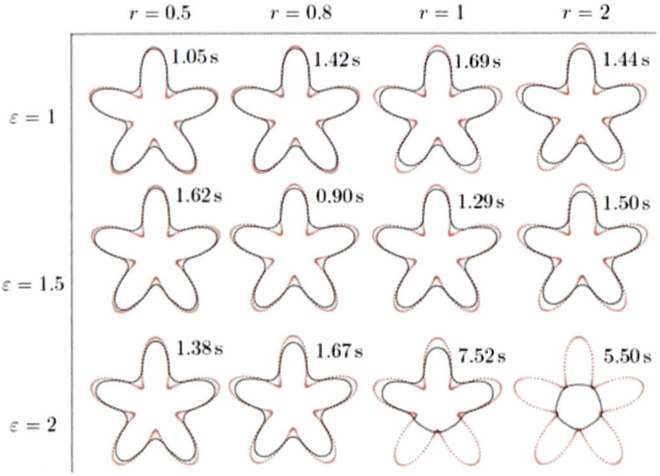


**Fig. 7** Results by ALM with different $r$ and $\varepsilon$. For each column, from top to bottom, $\varepsilon = 1, 1.5, 2$; and for each row, from left to right, $r = 0.5, 0.8, 1, 2$. Increasing $\varepsilon$ renders the curve less sharp and more convex. Increasing $r$ induces a stronger diffusion effect on $\phi^n$

$$\mathrm{Disc}\, Q^n = d^2 - 4(\phi^n)^4\pi^2|r\nabla\phi^n - \boldsymbol{\lambda}^{n-1}|^2.$$

The sign of $\phi^n(\mathbf{x})$ is related to the position of $\mathbf{x}$ relative to the 0-level-set. The sign of $\mathrm{Disc}\, Q^n$ is determined by comparing the length of a vector difference $r\nabla\phi^n - \boldsymbol{\lambda}^{n-1}$ with the quantity $d/(4(\phi^n)^2\pi)$. By the projection theorem, $|r\nabla\phi^n - \boldsymbol{\lambda}^{n-1}|^2$ is bounded below by $\alpha^n := |\boldsymbol{\lambda}^{n-1}|^2 - |\mathrm{Proj}_{\nabla\phi^n}\boldsymbol{\lambda}^{n-1}|^2 = |\boldsymbol{\lambda}^{n-1}|^2 - |\boldsymbol{\lambda}^{n-1}\cdot\nabla\phi^n|^2/(|\boldsymbol{\lambda}^{n-1}|^2|\nabla\phi^n|^2)$, i.e., the squared residual of orthogonal projection of $\boldsymbol{\lambda}^{n-1}$ onto $\nabla\phi^n$; therefore, we decide the sign of $\mathrm{Disc}\, Q^n$ using $r$ via the following cases:

1. When $\frac{d^2}{4(\phi^n)^4\pi^2} < \alpha^n$, for any $r > 0$, $\mathrm{Disc}\, Q^n < 0$.
2. When $\frac{d^2}{4(\phi^n)^4\pi^2} \geq \alpha^n$:
   (a) if $r > r_U^n$ or $r < r_L^n$, then $\mathrm{Disc}\, Q^n < 0$;
   (b) if $\max\{0, r_L^n\} \leq r \leq r_U^n$, then $\mathrm{Disc}\, Q^n \geq 0$.

   Here,

$$r_U^n = \frac{|\mathrm{Proj}_{\nabla\phi^n}\boldsymbol{\lambda}^{n-1}| + \sqrt{\frac{d^2}{4(\phi^n)^4\pi^2} - \alpha^n}}{|\nabla\phi^n|} \quad \text{and} \quad r_L^n = \frac{|\mathrm{Proj}_{\nabla\phi^n}\boldsymbol{\lambda}^{n-1}| - \sqrt{\frac{d^2}{4(\phi^n)^4\pi^2} - \alpha^n}}{|\nabla\phi^n|}.$$

When $\phi^n(\mathbf{x}) > 0$, $Q^n$ concaves upwards and $Q^n(0, r) \geq 0$ for any $r$. If Disc $Q^n < 0$, $Q^n$ is positive for all $\varepsilon$ and $d$ has no effect on level set evolution. If Disc $Q^n \geq 0$, $Q^n$ is positive for $\varepsilon$ outside the interval bounded by two roots of $Q^n$, i.e.,

$$0 < \varepsilon < \frac{d - \sqrt{\text{Disc } Q^n}}{2\phi^n \pi |r\nabla\phi^n - \boldsymbol{\lambda}^{n-1}|} \quad \text{or} \quad \varepsilon > \frac{d + \sqrt{\text{Disc } Q^n}}{2\phi^n \pi |r\nabla\phi^n - \boldsymbol{\lambda}^{n-1}|}.$$

When $\phi^n(\mathbf{x}) < 0$, $Q^n$ concaves downwards, and $Q^n(0, r) \leq 0$ for any $r$. In this case, $Q^n$ is never positive: either Disc $Q^n < 0$, i.e., no roots, or Disc $Q^n \geq 0$ but both roots are negative.

Notice that the bounds, $r_L^n$ and $r_U^n$, are closely related to the ratio $d/(\phi^n)^2$, which contributes to the adaptive behavior of ALM. For example, for a point $\mathbf{x}$ where $\phi^n(\mathbf{x}) > 0$, when $|\phi^n(\mathbf{x})|$ is close to 0 but $d(\mathbf{x}) \gg 0$, $r_L^n < 0$ and $r_U^n$ becomes extremely large; thus, for a moderate value of $r$, $d$ has a strong influence on the evolution of the level-set near $\mathbf{x}$ and swiftly moves the curve towards the point cloud. For a point $\mathbf{x}$ which is close to both $\mathcal{D}$ and $\{\phi^n = 0\}$, the level-set evolution becomes more stringent about the minimization of the energy (10).

Figure 8 illustrates this effect, for the five-fold circle point cloud in Fig. 1a with $r = 2$ and $\varepsilon = 1$. Figure 8 shows (a) Disc $Q^n$, (b) $r_U^n$, (c) $r_L^n$ , and (d) the region where $d$ effects the level set evolution. The figures are for iterations $n = 2, 3, 4, 7, 8, 10, 11, 13$ and 38 (converged). The region inside $\{\phi^n = 0\}$ always experiences the influence of $d$, as described above. Figure 8a shows that the region outside $\{\phi^n = 0\}$ is mostly blue indicating Disc $Q^n < 0$; hence, for almost every point outside the 0-level-set, as long as $r_L^n \leq r \leq r_U^n$, the landscape of $d$ has strong effects on the evolution. In (b) and (c), observe that high values of $r_U^n$ only concentrate near the 0-level-set while $r_L^n$ remains relatively small in the whole domain; thus, the influence of $d$ is strong near $\{\phi^n = 0\}$. (d) displays the white regions where $d$ explicitly guides the level-set evolution and the black regions where $d$ has no direct effect. These results show that, although ALM evolves the level-set globally, it ignores the effects of $d$ when evolving the regions far away from the level-sets; and it utilizes the values of $d$ to refine the local structures for the regions of the level-sets close to $\mathcal{D}$.

## 4 Conclusion

We propose two fast algorithms, SIM and ALM, to reconstruct a codimensional 1 submanifold from unstructured point clouds in $\mathbb{R}^2$ or $\mathbb{R}^3$ by minimizing the weighted minimum surface energy (2). SIM improves the computational efficiency by relaxing the constraint on the time-step using a semi-implicit scheme. ALM follows an augmented Lagrangian approach and solves the problem by an ADMM-type algorithm. Numerical experiments show that the proposed algorithms are superior in computational speed, and both of them produce accurate results. Theoretically, we

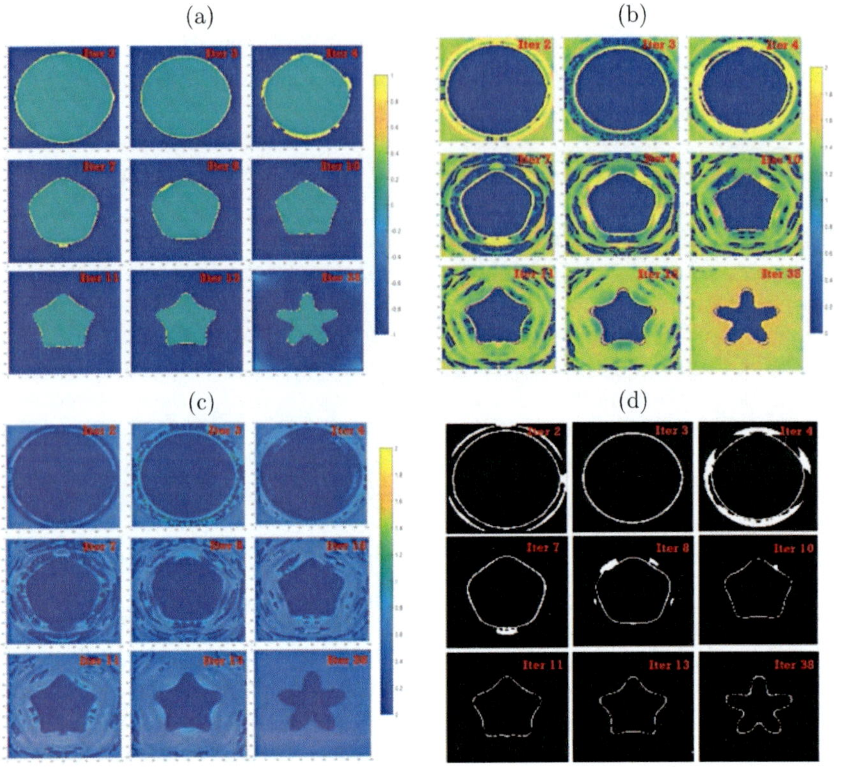


**Fig. 8** **a** Disc $Q^n$, **b** $r_U^n$, **c** $r_L^n$ at certain iterations. **d** The region (in white) where $d$ explicitly guides the level-set evolution by ALM. The distance function $d$ refines the local structures and it is only active near $\{\phi^n = 0\}$. This partially explains the efficiency of ALM

demonstrate the delicate interaction among parameters involved in ALM and show the connections between SIM and ALM. This explains the behaviors of ALM from the perspective of SIM.

## References

1. M. Alexa, J. Behr, D. Cohen-Or, S. Fleishman, D. Levin, C.T. Silva, Point set surfaces, in *Proceedings of the Conference on Visualization'01* (IEEE Computer Society, 2001), pp. 21–28
2. M. Alexa, J. Behr, D. Cohen-Or, S. Fleishman, D. Levin, C.T. Silva, Computing and rendering point set surfaces. IEEE Trans. Vis. Comput. Graph. **9**(1), 3–15 (2003)
3. E. Bae, X.-C. Tai, W. Zhu, Augmented Lagrangian method for an Euler's elastica based segmentation model that promotes convex contours. Inverse Probl. Imaging **11**(1), 1–23 (2017)
4. Z. Bi, L. Wang, Advances in 3D data acquisition and processing for industrial applications. Robot. Comput.-Integr. Manuf. **26**(5), 403–413 (2010)

5. M. Bolitho, M. Kazhdan, R. Burns, H. Hoppe, Parallel poisson surface reconstruction, in *International Symposium on Visual Computing* (Springer, 2009), pp. 678–689
6. R. Bracewell, R. Bracewell, *The Fourier Transform and Its Applications*. Electrical Engineering Series (McGraw Hill, 2000)
7. X. Bresson, S. Esedo?lu, P. Vandergheynst, J.-P. Thiran, S. Osher, Fast global minimization of the active contour/snake model. J. Math. Imaging Vis. **28**(2), 151–167 (2007)
8. F. Calakli, G. Taubin, SSD: smooth signed distance surface reconstruction, in *Computer Graphics Forum*, vol. 30 (Wiley Online Library, 2011), pp. 1993–2002
9. J.C. Carr, R.K. Beatson, J.B. Cherrie, T.J. Mitchell, W.R. Fright, B.C. McCallum, T.R. Evans, Reconstruction and representation of 3D objects with radial basis functions, in *Proceedings of the 28th Annual Conference on Computer Graphics and Interactive Techniques* (ACM, 2001), pp. 67–76
10. J.C. Carr, R.K. Beatson, B.C. McCallum, W.R. Fright, T.J. McLennan, T.J. Mitchell, Smooth surface reconstruction from noisy range data, in *Proceedings of the 1st International Conference on Computer Graphics and Interactive Techniques in Australasia and South East Asia* (ACM, 2003), pp. 119–ff
11. G. Casciola, D. Lazzaro, L.B. Montefusco, S. Morigi, Shape preserving surface reconstruction using locally anisotropic radial basis function interpolants. Comput. Math. Appl. **51**(8), 1185–1198 (2006)
12. T.F. Chan, L.A. Vese, Active contours without edges. IEEE Trans. Image Process. **10**(2), 266–277 (2001)
13. H.Q. Dinh, G. Turk, G. Slabaugh, Reconstructing surfaces using anisotropic basis functions, in *Proceedings Eighth IEEE International Conference on Computer Vision. ICCV 2001*, vol. 2 (IEEE, 2001), pp. 606–613
14. V. Estellers, M. Scott, K. Tew, S. Soatto, Robust poisson surface reconstruction, in *International Conference on Scale Space and Variational Methods in Computer Vision* (Springer, 2015), pp. 525–537
15. V. Estellers, D. Zosso, R. Lai, S. Osher, J.-P. Thiran, X. Bresson, Efficient algorithm for level set method preserving distance function. IEEE Trans. Image Process. **21**(12), 4722–4734 (2012)
16. L. Gomes, O.R.P. Bellon, L. Silva, 3D reconstruction methods for digital preservation of cultural heritage: a survey. Pattern Recogn. Lett. **50**, 3–14 (2014)
17. J. Haličková, K. Mikula, Level set method for surface reconstruction and its application in surveying. J. Surv. Eng. **142**(3), 04016007 (2016)
18. H. Huang, D. Li, H. Zhang, U. Ascher, D. Cohen-Or, Consolidation of unorganized point clouds for surface reconstruction. ACM Trans. Graph. (TOG) **28**(5), 176 (2009)
19. C.Y. Kao, S. Osher, J. Qian, Lax-Friedrichs sweeping scheme for static Hamilton-Jacobi equations. J. Comput. Phys. **196**(1), 367–391 (2004)
20. M. Kazhdan, M. Bolitho, H. Hoppe, Poisson surface reconstruction, in *Proceedings of the 4th Eurographics Symposium on Geometry Processing*, vol. 7, pp. 61–70 (2006)
21. M. Kazhdan, H. Hoppe, Screened poisson surface reconstruction. ACM Trans. Graph. (TOG) **32**(3), 29 (2013)
22. D. Khan, M.A. Shirazi, M.Y. Kim, Single shot laser speckle based 3D acquisition system for medical applications. Opt. Lasers Eng. **105**, 43–53 (2018)
23. H. Li, Y. Li, R. Yu, J. Sun, J. Kim, Surface reconstruction from unorganized points with $\ell_0$ gradient minimization. Comput. Vis. Image Underst. **169**, 108–118 (2018)
24. X. Li, W. Wan, X. Cheng, B. Cui, An improved Poisson surface reconstruction algorithm, in *2010 International Conference on Audio, Language and Image Processing* (IEEE, 2010), pp. 1134–1138
25. J. Liang, F. Park, H.-K. Zhao, Robust and efficient implicit surface reconstruction for point clouds based on convexified image segmentation. J. Sci. Comput. **54**(2–3), 577–602 (2013)
26. H. Liu, X. Wang, W. Qiang, Implicit surface reconstruction from 3D scattered points based on variational level set method, in *2008 2nd International Symposium on Systems and Control in Aerospace and Astronautics* (IEEE, 2008), pp. 1–5

27. H. Liu, Z. Yao, S. Leung, T.F. Chan, A level set based variational principal flow method for nonparametric dimension reduction on Riemannian manifolds. SIAM J. Sci. Comput. **39**(4), A1616–A1646 (2017)
28. S. Osher, R.P. Fedkiw, Level set methods: an overview and some recent results. J. Comput. Phys. **169**(2), 463–502 (2001)
29. S. Osher, J.A. Sethian, Fronts propagating with curvature-dependent speed: algorithms based on Hamilton-Jacobi formulations. J. Comput. Phys. **79**(1), 12–49 (1988)
30. A.C. Öztireli, G. Guennebaud, M. Gross, Feature preserving point set surfaces based on non-linear kernel regression, in *Computer Graphics Forum*, vol. 28 (Wiley Online Library, 2009), pp. 493–501
31. J.A. Sethian, *Level Set Methods and Fast Marching Methods: Evolving Interfaces in Computational Geometry, Fluid mechanics, Computer vision, and Materials Science*, vol. 3 (Cambridge University Press, 1999)
32. J. Shi, M. Wan, X.-C. Tai, D. Wang, Curvature minimization for surface reconstruction with features, in *International Conference on Scale Space and Variational Methods in Computer Vision* (Springer, 2011), pp. 495–507
33. Y. Shi, W.C. Karl, Shape reconstruction from unorganized points with a data-driven level set method, in *2004 IEEE International Conference on Acoustics, Speech, and Signal Processing*, vol. 3 (IEEE, 2004), pp. iii–13
34. P. Smereka, Semi-implicit level set methods for curvature and surface diffusion motion. J. Sci. Comput. **19**(1), 439–456 (2003)
35. X.-C. Tai, J. Hahn, G.J. Chung, A fast algorithm for Euler's elastica model using augmented Lagrangian method. SIAM J. Imaging Sci. **4**(1), 313–344 (2011)
36. A. Tsai, A. Yezzi Jr., W. Wells, C. Tempany, D. Tucker, A. Fan, W.E. Grimson, A. Willsky, A shape-based approach to the segmentation of medical imagery using level sets. IEEE Trans. Med. Imaging **22**(2), 137 (2003)
37. H.-K. Zhao, S. Osher, R. Fedkiw, Fast surface reconstruction using the level set method, in *Proceedings IEEE Workshop on Variational and Level Set Methods in Computer Vision* (IEEE, 2001), pp. 194–201
38. H.-K. Zhao, S. Osher, B. Merriman, M. Kang, Implicit, nonparametric shape reconstruction from unorganized points using a variational level set method. Comput. Vis. Image Underst. **80**(3), 295–319 (2000)

# A Total Variation Regularization Method for Inverse Source Problem with Uniform Noise

**Huan Pan and You-Wei Wen**

**Abstract** The problem of inverse source problem is considered in this paper. The main aim of this problem is to determine the source density function from the state function which is corrupted by uniform noise. Under the framework of maximum a posteriori estimator, the problem can be converted into an optimization problem where the objective function is composed of an $L_\infty$ norm and a total variation (TV) regularization term. By introducing an auxiliary variable, the optimization problem is further converted into a minimax problem. Then first order primal-dual method is applied to find the saddle point of the minimax problem. Numerical examples are given to demonstrate that our proposed method outperforms the other testing methods.

**Keywords** Inverse problem · Uniform noise · Total variation · $L_\infty$-norm constraint · Linear systems.

## 1 Introduction

In this paper, we consider the numerical solution of an elliptic inverse source problem [16, 17]. Inverse source problems arise in many areas of mathematical physics, and applications in recent year are rapidly expanding to such areas as geophysics, chemistry, medicine, engineering and mathematical imaging [5, 25]. The phenomena in these applications are generally described by partial differential equations. An inverse source problem for an elliptic partial differential equations on the domain $\Omega \in R^2$ with homogeneous Dirichlet boundary condition is given as follows [10]:

This work is supported by NSFC Grant No. 11871210, the Construct Program of the Key Discipline in Hunan Province, the SRF of Hunan Provincial Education Department (No.17A128)

H. Pan · Y.-W. Wen (✉)
Key Laboratory of Computing and Stochastic Mathematics (LCSM) (Ministry of Education of China), School of Mathematics and Statistics, Hunan Normal University, Changsha 410081, Hunan, P.R. China



X.-C. Tai et al. (eds.), *Mathematical Methods in Image Processing and Inverse Problems*, Springer Proceedings in Mathematics & Statistics 360,
https://doi.org/10.1007/978-981-16-2701-9_5

$$\begin{cases} -\nabla \cdot (a(x)\nabla u) + \langle b(x), \nabla u\rangle_{L^2(\Omega)} + c(x)u = f(x) & \text{in } \Omega \\ u = 0 & \text{on } \partial\Omega \end{cases} \tag{1}$$

where $a(x)$ and $c(x)$ are two given bounded and positive functions in $\Omega$, $b(x)$ denotes the velocity of flow, $u(x)$ represents state function, and $f(x)$ is the source density function. If the coefficients $a(x), c(x)$ and the source function $f(x)$ are directly given, we need to estimate the state function $u(x)$, the problem in (1) is called a forward source problem. However, in any physical and engineering problems such as pollutant detection and imaging science, we can acquire the state function $u(x)$ at the boundary of the reconstruction region, i.e., the measurement data $u(x)$ is available, but we need to estimate the source function $f(x)$. It is an inverse source problem [15]. The main aim of the inverse source problem is to determine $f$ from the state function $u$.

We shall focus our attention to find a numerical solution of the inverse source problem (1) in this paper. The discrete model of (1) can be represented by using vectors and matrices. With the lexicographical ordering of $\mathbf{u}$ and $\mathbf{f}$, their relationship can be expressed as follows:

$$K\mathbf{u} = \mathbf{f}.$$

Here $K$ is the matrix generated by the elliptic partial differential equations. Assume that the size of $\mathbf{u}$ is $N \times M$, $\mathbf{u}_{ij}$ denotes the $((i-1)N + j$-th component of $\mathbf{u}$. If the solution $\mathbf{u}$ is obtained, the source $\mathbf{f}$ can be computed directly by the matrix and the vector product. The solution $\mathbf{u}$ is generally associated with the boundary value which is an observation with errors, this is that $\mathbf{u}$ is corrupted by the noise $\mathbf{n}$ and the observation $\mathbf{u}_\delta$ is given by $\mathbf{u}_\delta = \mathbf{u} + \mathbf{n}$. Hence we obtain $\mathbf{f} = K(\mathbf{u}_\delta - \mathbf{n})$. Since the observation data is corrupted by the measurement errors (noise), the source $\mathbf{f}$ can not be calculated by the product of the matrix and the vector. The observation data $\mathbf{u}_\delta$ can be rewritten as

$$\mathbf{u}_\delta = K^{-1}\mathbf{f} + \mathbf{n}.$$

In mathematics, the inverse source problem is ill-conditioned in the sense of Hadamard [13, 20], namely, small perturbation (quantization errors) in the measurement data may lead to the lack of stability of numerical inversions. The ill-conditioning can be alleviated to stabilize the solution by incorporating the priori source information, and the solution $\mathbf{f}$ can be formulated as a minimizer of the following minimization problem

$$\min_{\mathbf{f}} \psi(\mathbf{u}_\delta, \mathbf{f}) + \lambda\phi(\mathbf{f}).$$

Here the function $\psi(\mathbf{u}_\delta, \mathbf{f})$ is the data-fitting term to represent the distribution of the measurement error $\mathbf{n}$, the function $\phi(\mathbf{f})$ is the regularization term to represent the prior knowledge of $\mathbf{f}$, and $\lambda$ is a regularization parameter.

In this paper, we assume that the measurement data $u(x)$ is corrupted by a uniform distribution noise. This is that $\mathbf{n}_i$ (the $i$-th entry of $\mathbf{n}$) are the independent identically distributed samples with uniform distribution $U(-c, c)$, here $c$ denotes the noise level. According the distribution function of $\mathbf{n}$, we can derive the data-fitting term by $\psi(\mathbf{u}_\delta, \mathbf{f}) = \left\| K^{-1}\mathbf{f} - \mathbf{u}_\delta \right\|_\infty$, see [10, 27]. In the literatures [2, 13, 18, 19], a Tikhonov-type function was used to represent the prior knowledge in the inverse problems.

Numerical difficulty is caused due to non-differentiability of the $L_\infty$-norm in the data-fitting term. In [10, 27], the minimization problem was reformulated into a constrained one. In [10], a Moreau-Yosida approximation for $L_\infty$-norm constraint was considered, and the authors then applied a semi-smooth Newton method to solve for the resulting optimality condition. In [27], the $L_\infty$-norm constraint was handled by active set constraints arising from the optimality conditions, and then an efficient semi-smooth Newton method was applied to find a solution.

In this paper, we consider that the source function is a piecewise continuous function and apply the total variation (TV) function [24] to represent its prior knowledge. The TV regularization has been widely used in many problems such as image denoising [1, 24], image restoration [3, 6], image segmentation [7, 8] and so on. However, to best of our knowledge, there are few papers using the TV function as a regularization term in the inverse source problem. We remark that both the data-fitting term and the regularization term considered in this paper are non-differentiable, we develop different numerical scheme to find a minimizer.

The remainder of the paper is structured as follows. In Sect. 2, we review the inverse source problem and propose total variation regularization method to find its solution. In Sect. 3, we transform the inverse source problem into an equivalent minimax problem and then apply first order primal-dual algorithm to solve it. In Sect. 4, Appling our proposed approach to address given numerical examples of the Inverse Source Problem. Finally, the Sect. 5 concludes this paper.

## 2 Total Variation Regularization for Inverse Source Problem

In this section, we consider a total variation (TV) regularization approach for inverse source problem. The minimization problem can be written as

$$\min_{\mathbf{f}} \left\| K^{-1}\mathbf{f} - \mathbf{u}_\delta \right\|_\infty + \lambda \left\| \nabla \mathbf{f} \right\|_1 . \tag{2}$$

Here $\|\nabla \mathbf{f}\|_1$ denotes the TV norm of $\mathbf{f}$. The TV norm is defined by $\phi(\mathbf{f}) = \|\nabla \mathbf{f}\|_1$, here

$$(\nabla \mathbf{f})_{i,j} = ((\nabla_x \mathbf{f})_{i,j}, (\nabla_y \mathbf{f})_{i,j})$$

with

$$(\nabla_x \mathbf{f})_{i,j} = \begin{cases} \mathbf{f}_{i+1,j} - \mathbf{f}_{i,j}, & \text{if } i < N, \\ 0, & \text{if } i = N, \end{cases} \quad (\nabla_y \mathbf{f})_{i,j} = \begin{cases} \mathbf{f}_{i,j+1} - \mathbf{f}_{i,j}, & \text{if } j < M, \\ 0, & \text{if } j = M. \end{cases}$$

We remark that the data-fitting term in (2) is derived by the assumption of uniform noise in the observation data. Considering an independent $U(-\delta, \delta)$ random variable $X$, where $\delta$ stands for the noise level. Since $\mathbf{n}_i$ (the $i$-th entry of $\mathbf{n}$) are the independent identically distributed samples with uniform distribution, the likelihood function is given by

$$\prod_{i=1}^{L} f_X(\mathbf{n}_i | \mathbf{u}_\delta, \delta) \propto \mathcal{I}(\mathbf{n}_1, \ldots, \mathbf{n}_L \in [-\delta, \delta]),$$

where the indicator function $\mathcal{I}(S)$ equals to 1 if $S$ happens and 0 otherwise. If at least one $\mathbf{n}_i$ (i.e., $(\mathbf{u}_\delta - K^{-1}\mathbf{f})_i$) falls outside of the interval $[-\delta, \delta]$, the likelihood will be equal to 0. Therefore, the solution of (2) should be any $\mathbf{u}$ that satisfies $\left\|\mathbf{u}_\delta - K^{-1}\mathbf{f}\right\|_\infty \leq \delta$. Therefore, the minimization problem in (2) can be rewritten as

$$\min_{\mathbf{f}} \|\nabla \mathbf{f}\|_1 \quad \text{s.t.} \quad \left\|K^{-1}\mathbf{f} - \mathbf{u}_\delta\right\|_\infty \leq \delta. \tag{3}$$

In fact, the minimization problem in (2) and (3) are mathematically equivalent. Given a regularization parameter $\lambda$ in (2), there exists a $\delta$ such that the solution of (2) is also the solution of (3). In contrast, given a $\delta$ in (3), there also exists a regularization parameter $\lambda$ in (2) such that the solution of (3) is also the solution of (2), moreover, $1/\lambda$ is the Lagrangian multiplier corresponding the $L_\infty$-norm inequality constraint. It is very important to choose a suitable regularization parameter $\lambda$ in (2), because $\lambda$ balances the data-fitting term and the regularization term and avoids to over-fitting or under-fitting the data. Compare to tune the regularization parameter $\lambda$, it is more easier to choose the noise level $\delta$ because $\delta$ is the noise level in the observation data. When $\delta$ is not available, it can be estimated by the method of moments [27]. In this paper, we will focus on the numerical scheme to solve (3). Although many methods have been proposed in the literature to find the minimizer of TV-based optimization problem, it is non-trivial to find the minimizer of (3) because both the TV norm and the $L_\infty$ norm are non-differentiable, also the minimizer should satisfy the inequality constraint. In the next section, we will consider the numerical scheme to find a minimizer of (3).

## 3 Primal-Dual Approach

In this section, we find the minimizer of the inverse source problem (3) by transforming it into a minimax problems. Then we solve it by a primal-dual method [9, 11, 14, 22, 23, 26, 30, 31]. We will apply Chambolle-Pock first order primal-dual algorithm in [9] to seek the saddle point of our minimax problem. We therefore give a brief introduction of the method here.

### 3.1 Chambolle-Pock's First-Order Primal-Dual Algorithm

In [9], Chambolle and Pock considered solving the minimax problem:

$$\min_{\mathbf{v}} \max_{\mathbf{z}} \Phi(\mathbf{v}) + \langle \mathbf{v}, H\mathbf{z} \rangle - \Psi(\mathbf{z}). \tag{4}$$

Here $\Phi$, $\Psi$ are propoer, convex and lower semi-continuous functions, and $H$ is a linear operator with induced norm $\|H\|$. They proposed to solve the problem by a first-order primal-dual algorithm as follows:

$$\begin{cases} \mathbf{v}^{(k+1)} = \underset{\mathbf{v}}{\operatorname{argmin}} \Phi(\mathbf{v}) + \langle \mathbf{v}, H\mathbf{z} \rangle + \frac{1}{2t} \left\| \mathbf{v} - \mathbf{v}^{(k)} \right\|_2^2, \\ \widehat{\mathbf{v}}^{(k+1)} = \mathbf{v}^{(k+1)} + \mu(\mathbf{v}^{(k+1)} - \mathbf{v}^{(k)}), \\ \mathbf{z}^{(k+1)} = \underset{\mathbf{z}}{\operatorname{argmax}} \left\langle \widehat{\mathbf{v}}^{(k+1)}, H\mathbf{z} \right\rangle - \Psi(\mathbf{z}) - \frac{1}{2s} \left\| \mathbf{z} - \mathbf{z}^{(k)} \right\|_2^2. \end{cases} \tag{5}$$

The parameters $s, t > 0$ are step sizes of the primal and dual variables respectively, and $\mu$ is the combination parameter. In the iterative procedure, proximal-point iterations are applied to the sub-differentials of the $\mathbf{v}$ and $\mathbf{z}$ subproblems in (5) with the primal variable and the dual variable fixed alternately.

### 3.2 Minimax Problem

Let us describe the notations that we will use in the followings. For $\boldsymbol{\xi} \in \mathbb{R}^{NM} \times \mathbb{R}^{NM}$, $\boldsymbol{\xi}_{i,j} = (\boldsymbol{\xi}_{i,j,1}, \boldsymbol{\xi}_{i,j,2}) \in \mathbb{R}^2$ denotes the $(i + (j-1)n)$-th component of $\boldsymbol{\xi}$. Define the inner product $\langle \boldsymbol{\xi}, \mathbf{q} \rangle = \sum_{i,j} \boldsymbol{\xi}_{i,j} \mathbf{q}_{i,j}$ for $\boldsymbol{\xi}, \mathbf{q} \in \mathbb{R}^{nm} \times \mathbb{R}^{nm}$. Define $\|\boldsymbol{\xi}\|_\infty = \max_{i,j} |\boldsymbol{\xi}_{i,j}|$. Define $\operatorname{div} = -\nabla^T$ as the discrete version of the divergence operator, where $\nabla^T$ is the adjoint of $\nabla$, i.e.,

$$(\operatorname{div} \boldsymbol{\xi})_{i,j} = \begin{cases} \xi^x_{i,j} & i = 1 \\ \xi^x_{i,j} - \xi^x_{i-1,j} & 1 < i < N \\ -\xi^x_{i-1,j} & i = N \end{cases} + \begin{cases} \xi^y_{i,j} & j = 1, \\ \xi^y_{i,j} - \xi^y_{i,j-1} & 1 < j < N, \\ -\xi^y_{i,j-1} & j = N. \end{cases}$$

We represent the TV norm using the dual form, i.e.,

$$\|\nabla \mathbf{f}\|_1 = \max_{\|\boldsymbol{\xi}\|_\infty \le 1} \langle \operatorname{div} \boldsymbol{\xi}, \mathbf{f} \rangle. \tag{6}$$

Using the dual formulation, the minimization problem (3) can be written as the following minimax problem:

$$\min_{\left\|K^{-1}\mathbf{f} - \mathbf{u}_\delta\right\|_\infty \le \delta} \max_{\|\boldsymbol{\xi}\|_\infty \le 1} \langle \mathbf{f}, \operatorname{div} \boldsymbol{\xi} \rangle. \tag{7}$$

Introducing the auxiliary varible $\mathbf{r} = \mathbf{u}_\delta - K^{-1}\mathbf{f}$, we obtain $\mathbf{f} - K(\mathbf{u}_\delta - \mathbf{r}) = 0$. We consider Lagrangian function for the resulting equation

$$\mathcal{L}(\mathbf{f}, \mathbf{r}, \boldsymbol{\xi}, \mathbf{y}) \equiv \langle \mathbf{f}, \mathrm{div}\boldsymbol{\xi} \rangle + \langle \mathbf{y}, \mathbf{f} - K(\mathbf{u}_\delta - \mathbf{r}) \rangle . \tag{8}$$

Here $\mathbf{y}$ is the Lagrange multiplier associated with the equality constraint $\mathbf{f} - K(\mathbf{u}_\delta - \mathbf{r}) = 0$. Hence, we have

$$\max_{\|\boldsymbol{\xi}\|_\infty \le 1, \mathbf{y}} \mathcal{L}(\mathbf{f}, \mathbf{r}, \boldsymbol{\xi}, \mathbf{y}) = \begin{cases} \|\nabla \mathbf{f}\|_1 , & \text{if } \mathbf{f} - K(\mathbf{u}_\delta - \mathbf{r}) = 0, \\ \infty, & \text{otherwise.} \end{cases}$$

Also we have

$$\min_{\mathbf{f}} \mathcal{L}(\mathbf{f}, \mathbf{r}, \boldsymbol{\xi}, \mathbf{y}) = \begin{cases} \langle \mathrm{div}\boldsymbol{\xi}, K(\mathbf{u}_\delta - \mathbf{r}) \rangle , & \text{if } \mathrm{div}\boldsymbol{\xi} + \mathbf{y} = 0, \\ -\infty, & \text{otherwise.} \end{cases}$$

According to [4, Proposition 5.5.4], we know that the minimum and the maximum in (8) can be swapped and there exists a saddle point of $\mathcal{L}$. We obtain

$$\min_{\|\mathbf{r}\|_\infty \le \delta, \mathbf{f}} \max_{\|\boldsymbol{\xi}\|_\infty \le 1, \mathbf{y}} \mathcal{L}(\mathbf{f}, \mathbf{r}, \boldsymbol{\xi}, \mathbf{y}) = \max_{\|\boldsymbol{\xi}\|_\infty \le 1} \min_{\|\mathbf{r}\|_\infty \le \delta} \langle \mathrm{div}\boldsymbol{\xi}, K(\mathbf{u}_\delta - \mathbf{r}) \rangle .$$

Thus we have the following theorem.

**Theorem 1** *Define* $\mathcal{Q}(\mathbf{r}, \boldsymbol{\xi}) = \langle \mathrm{div}\boldsymbol{\xi}, K(\mathbf{u}_\delta - \mathbf{r}) \rangle$, *then we have*

$$\min_{\left\| K^{-1}\mathbf{f} - \mathbf{u}_\delta \right\|_\infty \le \delta} \|\nabla \mathbf{f}\|_1 = \max_{\|\boldsymbol{\xi}\|_\infty \le 1} \min_{\|\mathbf{r}\|_\infty \le \delta} \mathcal{Q}(\mathbf{r}, \boldsymbol{\xi}).$$

*Moreover, the minimum in the left-hand side above is attained at* $\mathbf{f}^* = K(\mathbf{u}_\delta - \mathbf{r}^*)$, *here* $(\mathbf{r}^*, \boldsymbol{\xi}^*)$ *is the saddle point of the function* $\mathcal{Q}(\mathbf{r}, \boldsymbol{\xi})$.

Now we apply Chambolle-Pock's first-order primal-dual method (5) to compute the saddle point of $\mathcal{Q}(\mathbf{r}, \boldsymbol{\xi})$, the iterative scheme is given as follows:

$$\mathbf{r}^{k+1} = \operatorname*{argmin}_{\|\mathbf{r}\|_\infty \le \delta} \mathcal{Q}(\mathbf{r}, \boldsymbol{\xi}^k) + \frac{1}{2s} \left\| \mathbf{r} - \mathbf{r}^k \right\|_2^2 \tag{9}$$

$$\widehat{\mathbf{r}}^{k+1} = \mathbf{r}^{k+1} + \theta(\mathbf{r}^{k+1} - \mathbf{r}^k) \tag{10}$$

$$\boldsymbol{\xi}^{k+1} = \operatorname*{argmax}_{\|\boldsymbol{\xi}\|_\infty \le 1} \mathcal{Q}(\widehat{\mathbf{r}}^{k+1}, \boldsymbol{\xi}) - \frac{1}{2t} \left\| \boldsymbol{\xi} - \boldsymbol{\xi}^k \right\|_2^2 \tag{11}$$

### 3.3 Subproblem for $\mathbf{r}$

The minimization of (9) reduces to

$$\mathbf{r}^{k+1} = \underset{\mathbf{r}}{\operatorname{argmin}} \left\langle \operatorname{div}\boldsymbol{\xi}^k, K(\mathbf{u}^\delta - \mathbf{r})\right\rangle + \frac{1}{2s}\left\|\mathbf{r} - \mathbf{r}^k\right\|_2^2 \tag{12}$$

$$= \underset{\mathbf{r}}{\operatorname{argmin}} \left\|\mathbf{r} - (\mathbf{r}^k - sK^T\operatorname{div}\boldsymbol{\xi}^k)\right\|_2^2 \tag{13}$$

We first introduce the concept of the projection operator.

$$\mathcal{P}(\mathbf{w}) = \underset{\mathbf{r}\in\Omega}{\operatorname{argmin}} \|\mathbf{r} - \mathbf{w}\|_2^2 . \tag{14}$$

In general, the projection onto a general convex set is difficult and computationally expensive. As the $L_\infty$-constraints can be formulated as the bounded constraints, the corresponding closed-form solution is given by

$$[\mathcal{P}(\mathbf{w})]_i = \begin{cases} \delta, & \mathbf{w}_i \geq \delta. \\ \mathbf{w}_i, & |\mathbf{w}_i| < \delta. \\ -\delta, & \mathbf{w}_i \leq -\delta. \end{cases}$$

By using a suitable projection operator, we can view $\mathbf{r}^{k+1}$ as the projection of $\left(\mathbf{r}^{k+1} - sK^T\operatorname{div}\boldsymbol{\xi}^k\right)$ on $\Omega$. Thus we obtain

$$\mathbf{r}^{k+1} = \mathcal{P}\left(\mathbf{r}^{k+1} - sK^T\operatorname{div}\boldsymbol{\xi}^k\right). \tag{15}$$

### 3.4 Subproblem for $\boldsymbol{\xi}$

We change the maximization problem for $\boldsymbol{\xi}$ in (11) to a minimization one and obtain:

$$\boldsymbol{\xi}^{k+1} = \underset{\boldsymbol{\xi}}{\operatorname{argmax}} \left\langle \operatorname{div}\boldsymbol{\xi}, K(\mathbf{u}^\delta - \widehat{\mathbf{r}}^{k+1})\right\rangle - \frac{1}{2t}\left\|\boldsymbol{\xi} - \boldsymbol{\xi}^k\right\|_2^2 \tag{16}$$

$$= \underset{\|\boldsymbol{\xi}\|_\infty\leq 1}{\operatorname{argmin}} -\left\langle \operatorname{div}\boldsymbol{\xi}, K(\mathbf{u}^\delta - \widehat{\mathbf{r}}^{k+1})\right\rangle + \frac{1}{2t}\left\|\boldsymbol{\xi} - \boldsymbol{\xi}^k\right\|_2^2 \tag{17}$$

Thus

$$\boldsymbol{\xi}^{k+1} = \mathcal{P}_{\mathcal{A}}(\boldsymbol{\xi}^k - t\nabla K(\mathbf{u}^\delta - \widehat{\mathbf{r}}^{k+1}))$$

where $\mathcal{A} = \left\{\boldsymbol{\xi} : \|\boldsymbol{\xi}\|_\infty \leq 1\right\}$, the gradient projection of $(\boldsymbol{\xi}^k - t\nabla\mathbf{f}^{k+1})$ onto the set $\mathcal{A}$. In the following, we derive a formula for the gradient projection operator

$$\mathcal{P}_{\mathcal{A}}(\mathbf{q}) = \underset{\mathbf{p}\in\mathcal{A}}{\text{argmin}}\, \|\mathbf{p} - \mathbf{q}\|_2^2$$

For any $\mathbf{q}$, by the definition of the set $\mathcal{A}$, the Lagrangian function is

$$\|\mathbf{p} - \mathbf{q}\|_2^2 + \sum_{i,j} t_{i,j}\left(|p_{i,j}|^2 - 1\right),$$

where $t_{i,j} \geq 0$ is the Lagrangian multiplier associated with the constraint $|p_{i,j}|^2 \leq 1$. Its complementarity conditions implies that for the optimal $t_{i,j}$, either $t_{i,j} = 0$ with $|p_{i,j}|, |q_{i,j}| < 1$, or $t_{i,j} > 0$ with $|p_{i,j}| = 1$ and $|q_{i,j}| \geq 1$. In the former case, we have $p_{i,j} = q_{i,j}$. In the latter case, the KKT conditions yields $p_{i,j} - q_{i,j} + t_{i,j}p_{i,j} = 0$ for all $i, j$. Therefore, we have $t_{i,j} = |q_{i,j}| - 1$, and thus $p_{i,j} = q_{i,j}/|q_{i,j}|$. Hence, we obtain

$$(\mathcal{P}_{\mathcal{A}}(\mathbf{q}))_{i,j} = \frac{1}{\max(1, |q_{i,j}|)} q_{i,j}. \tag{18}$$

## 4 Numerical Results

In this section, three numerical experiments are implemented to demonstrate the effectiveness of the proposed method, that is to consider the inverse source problem (1) with domain $\Omega = [0, 1]^2$. We investigate the influence of noise level on the numerical results, specifically, set $\delta = 0.1, 0.2, 0.3, 0.4, 0.5, 0.6, 0.7, 0.8, 0.9$. In order to be able to stabilize the convergence of approximate solution by our proposed method, choosing primal variation step size $s = 5 \times 10^{-8}$ and dual variation step size $t = 5 \times 10^{-2}$. We show the exact solution $\mathbf{f}^*$ with size $256 \times 256$ in Fig. 1a, the exact data $\mathbf{u}^*$ in Fig. 1b and observation data $\mathbf{u}_\delta$ with noise level $\delta = 0.1, 0.4, 0.9$ respectively in Fig. 2.

In the following experiments, we compare our algorithm(TV) with semi-smooth Newton mehthod (SSN) [27], Primal-Dual method (PD) [21] and Forward Backward

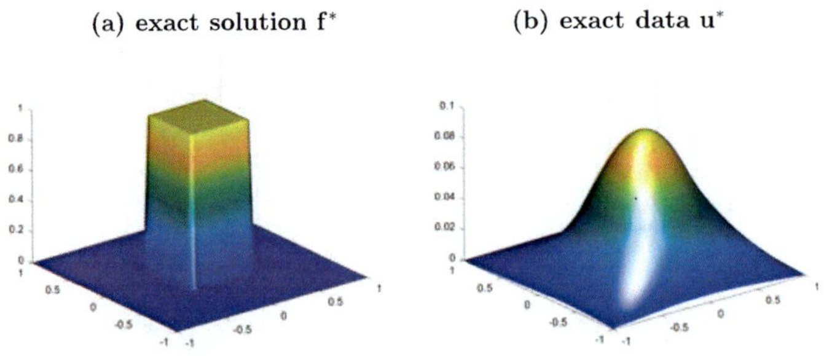


**Fig. 1** Left: the exact solution $\mathbf{f}^*$ with size $256 \times 256$; Right: the exact data $\mathbf{u}^*$

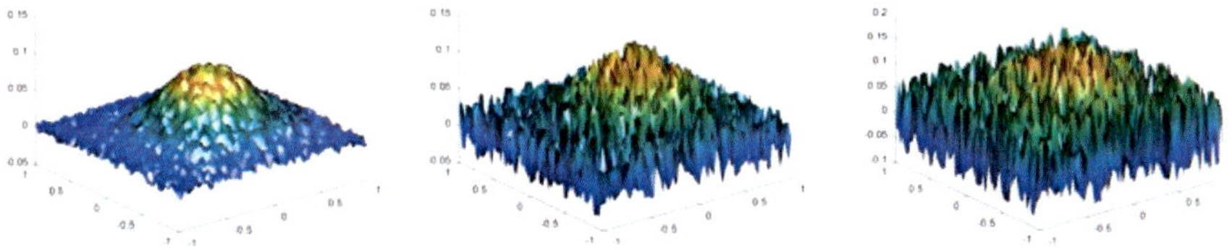

$\delta = 0.1$ $\delta = 0.4$ $\delta = 0.9$

**Fig. 2** Observed data $\mathbf{u}_\delta$ corrupted by uniform noise levels with the $\delta$ = 0.1, 0.4, 0.9 respectively

method (FB) [12]. The Root-Mean-Square-Error (RMSE) is used to quantitatively measure the quality of the estimated solution. It is defined as follows:

$$\text{RMSE} = \frac{1}{\sqrt{L}} \left\| \widehat{\mathbf{f}} - \mathbf{f}^* \right\|_2$$

where $\mathbf{f}^*$ denotes the exact solution and $\widehat{\mathbf{f}}$ denotes the estimated solution. The smaller RMSE is, the better the estimated solution is.

We consider the discrete problem (1) and set $a(x) = 1$ and $c(x) = 0$. Three different functions for $b(x)$ are used in the tests, they are $b(x) = -[2, 0]$, $b(x) = -[0, 1]$ and $b(x) = -[2, 1]$ respectively. In order to quantitatively measure the accuracy of the estimated solutions, we show the RMSE values for different noise level in Table 1. We note that the RMSE of recovery data by four methods gradually increase with the noise level increasing. The RMSE obtained by TV method is smaller than that obtained by other methods.

We show the estimated solutions obtained by different methods in the Figs. 3, 4 and 5 with different noise levels $\delta = 0.1, 0.4, 0.9$, respectively. We can observe that there are some jumps in the estimated solutions obtained by SNN method and FB method. The estimated solutions obtained by PD method look smooth. We remark that the Tikhonov-type regularization function is applied in the these three methods. It is obviously that the estimated solutions obtained by the proposed method are closer to the true solution.

## 5 Conclusion

In this paper, we study the inverse source problem where observation data are corrupted by uniform noise. The main contribution of this paper is to develop an efficient total variation regularization method for solving the ill-posed inverse source problem of the $L_\infty$-norm data fitting. Numerical examples are given to demonstrate that our proposed method outperforms the other testing methods.

**Table 1** RMSE of estimated solution for different noise levels

| $\delta$ | $\|\mathbf{u}_\delta - \mathbf{u}\|_2$ | SSN | PD | FB | TV |
|---|---|---|---|---|---|
| | | $a = 1, b = -[2, 0]$ | | | |
| 0.1 | 9.36e-03 | 1.19e-01 | 1.25e-01 | 1.34e-01 | 2.36e-02 |
| 0.3 | 2.80e-02 | 1.33e-01 | 1.36e-01 | 1.39e-01 | 3.85e-02 |
| 0.5 | 4.68e-02 | 1.33e-01 | 1.41e-01 | 1.48e-01 | 6.95e-02 |
| 0.7 | 6.55e-02 | 1.49c-01 | 1.46e-01 | 1.53e-01 | 9.94e-02 |
| 0.9 | 8.42e-02 | 1.53e-01 | 1.50e-01 | 1.58e-01 | 1.18e-01 |
| | | $a = 1, b = -[0, 1]$ | | | |
| 0.1 | 9.93e-03 | 1.19e-01 | 1.26e-01 | 1.35e-01 | 1.82e-02 |
| 0.3 | 2.97e-02 | 1.32e-01 | 1.36e-01 | 1.41e-01 | 3.34e-02 |
| 0.5 | 4.96e-02 | 1.34e-01 | 1.42e-01 | 1.47e-01 | 7.16e-02 |
| 0.7 | 6.95e-02 | 1.50e-01 | 1.47e-01 | 1.55e-01 | 9.73e-02 |
| 0.9 | 8.93e-02 | 1.52e-01 | 1.51e-01 | 1.60e-01 | 8.86e-02 |
| | | $a = 1, b = -[2, 1]$ | | | |
| 0.1 | 9.24e-03 | 1.18e-01 | 1.25e-01 | 1.27e-01 | 1.95e-02 |
| 0.3 | 2.77e-02 | 1.32e-01 | 1.35e-01 | 1.37e-01 | 3.65e-02 |
| 0.5 | 4.62e-02 | 1.32e-01 | 1.41e-0 | 1.43e-01 | 7.03e-02 |
| 0.7 | 6.47e-02 | 1.50e-01 | 1.47e-01 | 1.50e-01 | 1.04e-01 |
| 0.9 | 8.32e-02 | 1.51e-01 | 1.49e-01 | 1.52e-01 | 1.16e-01 |

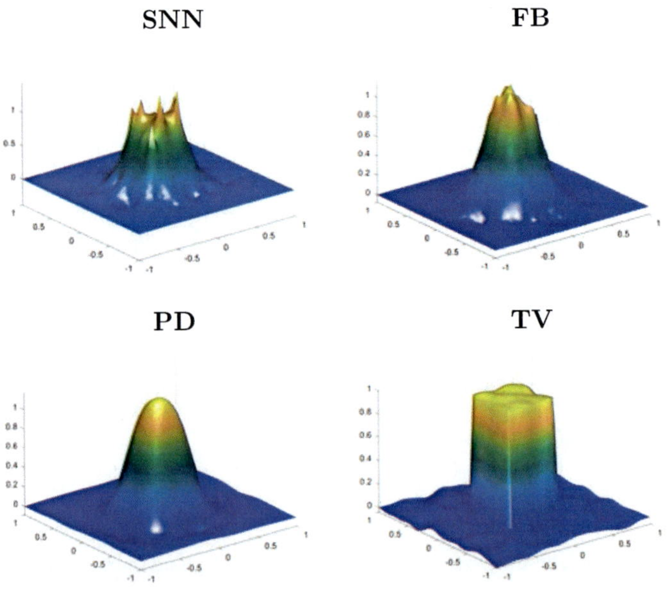


**Fig. 3** Root-mean-square-error (RMSE) values obtained by different methods for different noise levels. Here $\delta = 0.1$ and $a = 1, b = -[2, 0]$

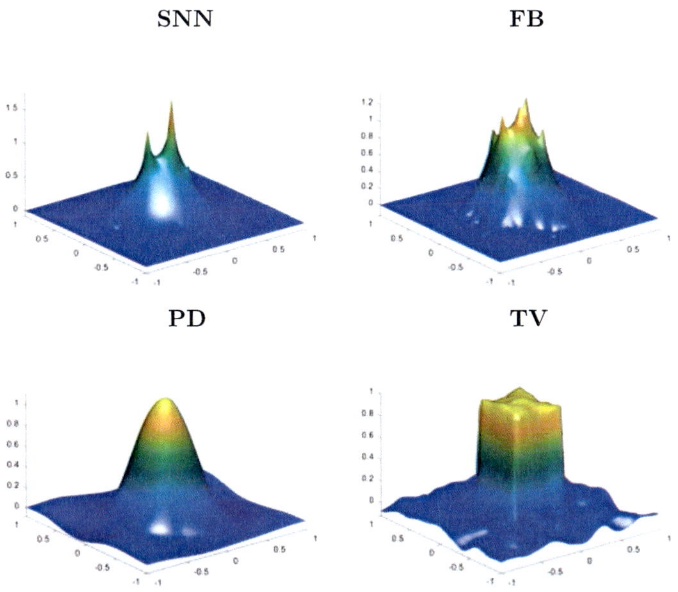


**Fig. 4** Four algorithms recovering data graphs with noise levels of $d = 0.4$ based on $a = 1, b = -[0, 1]$

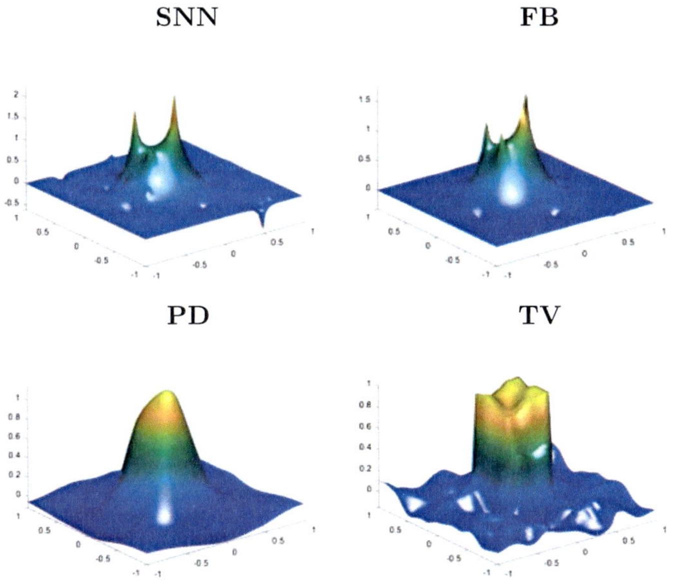


**Fig. 5** Four algorithms recovering data graphs with noise levels of $\delta = 0.9$ based on $a = 1, b = -[2, 1]$

## References

1. J. Aujol, G. Gilboa, Constrained and SNR-based solutions for TV-Hilbert space image denoising. J. Math. Imaging Vis. **26**(1), 217–237 (2006)
2. V. Akcelik, G. Biros, O. Ghattas, K. Long, B.G.V. Bloemen Waanders, A variational finite element method for source inversion for convective-diffusive transport. Finite Elem. Anal. Des. **39**(8), 683–705 (2003)
3. M. Bertalmio, V. Caselles, B. Rougé, A. Solé, TV based image restoration with local constraints. J. Sci. Comput. **19**(1–3), 95–122 (2003)
4. D. Bertsekas, *Convex Optimization Theory* (Athena Scientific Belmont, MA, 2009)
5. A. Badia, T. Ha-Duong, An inverse source problem in potential analysis. Inverse Probl. (2000)
6. P. Blomgren, T. Chan, Color TV: total variation methods for restoration of vector-valued images. IEEE Trans. Image Process. **7**(3), 304–309 (1998)
7. X. Cai, R. Chan, M. Nikolova, T. Zeng, A three-stage approach for segmenting degraded color images: Smoothing, lifting and Thresholding (SlaT). J. Sci. Comput. **72**(3), 1313–1332 (2017)
8. X. Cai, R. Chan, T. Zeng, A two-stage image segmentation method using a convex variant of the Mumford-Shah model and thresholding. SIAM J. Imaging Sci. **6**(1), 368–390 (2013)
9. A. Chambolle, T. Pock, A first-order primal-dual algorithm for convex problems with applications to imaging. J. Math. Imaging Vis. **40**(1), 120–145 (2011)
10. C. Clason, $L_\infty$ fitting for inverse problems with uniform noise. Inverse Probl **28**(10) (2012)
11. G. Chen, M. Teboulle, A proximal-based decomposition method for convex minimization problems. Math. Program. Ser. A **64**(1):81–101 (1994)
12. P. Combettes, V. Wajs, Signal recovery by proximal forward-backward splitting . Multiscale Model. Simul. **4**(4), 1168–1200 (2005)
13. H. Engl, R. Ramlau, *Regularization of Inverse Problems*, Encyclopedia of Applied and Computational Mathematics (Springer, Berlin, Heidelberg, 2015)
14. J. Eckstein, D. Bertsekas, On the Douglas-Rachford splitting method and the proximal point algorithm for maximal monotone operators. Math. Program. Ser. A **55**(3), 293–318 (1992)
15. Q. Hu, S. Shu, J. Zou, A new variational approach for inverse source problems. Numer. Math.-Theory Methods Appl. **12**(2), 331–347 (2019)
16. V. Isakov, Inverse source problems. Ams Ebooks Prog. **34**, 191 (1990)
17. V. Isakov, Inverse problems for partial differential equations. Appl. Math. Sci. **703**(45), 93–98 (1979)
18. Y. Keung, J. Zou, Numerical identifications of parameters in parabolic systems. Inverse Probl. **14**(1), 83–100 (1998)
19. Y. Keung, J. Zou, X. Wang, An efficient linear solver for nonlinear parameter identification problems. J. Sci. Comput. (1998)
20. E. Lavrent, M. Jn, et al., Inverse Probl. Math. Phys. (1987)
21. X. Liu, Z. Chen, Y. Wen, A dual method for uniform noise removal base on $L_\infty$ norm constraint, pp. 1346–1350, 07 (2017)
22. R. Rockafellar, Augmented Lagrangians and applications of the proximal point algorithm in convex programming. Math. Oper. Res. **1**(2), 97–116 (1976)
23. R. Rockafellar, Monotone operators and the proximal point algorithm. SIAM J. Control Optim. **14**(5), 877–898 (1976)
24. L. Rudin, S. Osher, E. Fatemi, Nonlinear total variation based noise removal algorithms. Physica D **60**, 259–268 (1992)
25. A. Tikhonov, A. Goncharsky, V. Stepanov. *Numerical Methods for the Solution of Ill-Posed Problems* (Kluwer Academic Publishers, 1995)
26. P. Tseng, Applications of a splitting algorithm to decomposition in convex programming and variational inequalities. SIAM J. Control Optim. **29**(1), 119–138 (1991)
27. Y. Wen, W. Ching, M. Ng, A semi-smooth newton method for inverse problem with uniform noise. J. Sci. Comput. **75**(2), 713–732 (2018)
28. Y. Yang, N. Galatsanos, A. Katsaggelos, Projection-based spatially adaptive reconstruction of block-transform compressed images. IEEE Trans. Image Process. **4**(7), 896–908 (1995)

29. L. Zhen, E. Delp, Block artifact reduction using a transform-domain Markov random field model. IEEE Trans. Circuits Syst. Video Technol. **15**(12), 1583–1593 (2005)
30. M. Zhu, *Fast Numerical Algorithms for Total Variation Based Image Restoration*. Ph.D. thesis, University of California, Los Angeles (2008)
31. M. Zhu, T. Chan, An efficient primal-dual hybrid gradient algorithm for total variation image restoration. *UCLA CAM Report*, pp. 08–34 (2007)

# Automatic Parameter Selection Based on Residual Whiteness for Convex Non-convex Variational Restoration

**Alessandro Lanza, Serena Morigi, and Fiorella Sgallari**

**Abstract** Image restoration is a well-known ill-posed inverse problem whose aim is to recover a sharp clean image from the corresponding blur- and noise-corrupted observation. Variational methods penalize solutions deemed undesirable by incorporating regularization techniques. A popular strategy relies on using sparsity promoting regularizers; it is well known that, in general, nonconvex regularizers hold the potential for promoting sparsity more effectively than convex regularizers. Recently a new class of convex non-convex (CNC) variational models has been proposed which includes a general parametric nonconvex nonseparable regularizer. However, the performance of this approach depends critically on the regularization parameter. In this paper we propose to use a parametric CNC variational restoration model within a bilevel framework, where the parameter is tuned by minimizing a measure of the restoration residual whiteness. Some preliminary numerical experiments are shown which indicate the effectiveness of the proposal.

**Keywords** Sparsity-inducing regularization · Variational methods · Ill-posed problems · Non-convex non-smooth regularization · Optimization · Additive white gaussian noise.

---

Dedicated to Raymond H. Chan on the occasion of his 60th birthday.

---

A. Lanza · S. Morigi (✉) · F. Sgallari
Department of Mathematics, University of Bologna, Bologna, Italy
e-mail: serena.morigi@unibo.it

A. Lanza
e-mail: alessandro.lanza2@unibo.it

F. Sgallari
e-mail: fiorella.sgallari@unibo.it



X.-C. Tai et al. (eds.), *Mathematical Methods in Image Processing and Inverse Problems*,
Springer Proceedings in Mathematics & Statistics 360,
https://doi.org/10.1007/978-981-16-2701-9_6

# 1 Introduction

In this paper, we consider the problem of restoring 2-D gray-scale images corrupted by blur and additive white Gaussian noise (AWGN).

These images can be represented by the discretization of a real valued function defined on a 2-D compact rectangular domain. Let $x \in \mathbb{R}^n$, with $n = n_1 n_2$, be the unknown $n_1 \times n_2$ clean image concatenated into an $n$-vector, $A \in \mathbb{R}^{n \times n}$ be a known blurring operator and $\epsilon \in \mathbb{R}^n$ be an unknown realization of the noise process, which we assume white Gaussian with zero-mean and standard deviation $\sigma$. The discrete imaging model of the degradation process which relates the observed degraded image $b \in \mathbb{R}^n$ with $x$, can be expressed as follows:

$$b = Ax + \epsilon \,. \tag{1}$$

Given $A$ and $b$, our goal is to solve the inverse problem of recovering an accurate estimate of $x$, which is known as deconvolution or deblurring. When $A$ is the identity operator, recovering $x$ is referred as denoising.

Image deblurring is a discrete ill-posed problem, as such further a priori assumptions on the solution can help to determine a meaningful approximation of $x$. Assuming the image is corrupted by AWGN, then an estimate $x_\lambda^*$ of $x$ can be obtained as a solution—i.e., a global minimizer—of the following variational model which is the sum of a convex smooth (quadratic) fidelity term and a regularization term:

$$x_\lambda^* \in \arg\min_{x \in \mathbb{R}^n} \mathcal{J}(x;\lambda), \qquad \mathcal{J}(x;\lambda) := \frac{1}{2}\,\|Ax - b\|_2^2 + \lambda\,\mathcal{R}(x)\,, \tag{2}$$

where $\|v\|_2$ denotes the $\ell_2$ norm of vector $v$ and $\lambda$ represents the classical regularization parameter which controls the trade-off between data-fidelity and regularization.

The regularizer $\mathcal{R}(x)$ encodes a priori knowledge on the solution. Focusing on the recovery of images characterized by some sparsity property, we consider the general class of sparsity-inducing variational models described in [15] to determine solutions $x_\lambda^*$ which are close to the data $b$ according to the observation model and, at the same time, for which the transformed solution vector $y_\lambda^* = G(Lx_\lambda^*)$ is sparse with $L \in \mathbb{R}^{r \times n}$ a linear operator and $G : \mathbb{R}^r \to \mathbb{R}^s$ a possibly nonlinear vector-valued function—see [15].

The drawback of the proposal in [15], which will be briefly illustrated in Sect. 3, is that it requires a trial-and-error procedure for tuning the regularization parameter $\lambda$ and a manual stopping. This represents a crucial aspect in variational restoration methods and has been subject of several research works.

We propose an automatic criterion for adjusting the regularization parameter $\lambda$. More precisely, our proposal is based on the key idea that if the restored image is a good estimate of the target clean image, then the residual image must resemble the realization of the noise process, thus being spectrally white.

Hence, starting from a sufficiently small $\lambda$ value, we iteratively increase $\lambda$ until a suitable whiteness maximality criterion is satisfied.

In Sect. 2 we review some related works on the choice of the regularization parameter. The class of CNC variational models introduced in [15] is briefly illustrated in Sect. 3. In Sect. 4 we define the residual whiteness strategy, and Sect. 5 is devoted to the description of the proposed algorithmic framework. Numerical results are presented in Sect. 6. Conclusions are drawn in Sect. 7.

## 2 Related Work

A crucial issue in the regularization of ill-posed inverse problems is the choice of the regularization parameter. The quality of the solution is affected by the value of $\lambda$: a too large value of $\lambda$ gives an over-smoothed solution that lacks details that the desired original solution may have, while a too small value of $\lambda$ yields a computed solution that is unnecessarily, and possibly severely, contaminated by propagated error that stems from the error $\epsilon$ in $b$.

The discrepancy principle (DP) [22] chooses the regularization parameter so that the variance of the residual equals that of the noise; the DP thus requires an accurate estimate of the noise variance and is known to yield overregularized estimates [7]. The sensitivity of $\lambda$ and of the computed solution to the inaccuracies in an available estimate of $\|\epsilon\|$ has been investigated by Hamarik et al. [6], who proposed alternatives to the discrepancy principle when only a poor estimate of $\|\epsilon\|$ is known. Automatic procedures for selecting the $\lambda$ parameter based on the DP has been proposed in literature, see e.g. [9].

Parameter choice methods when no estimate of $\|\epsilon\|$ is available are commonly referred to as "heuristic", because they may fail in certain situations; see [5].

A large number of heuristic parameter choice methods have been proposed in the literature due to the importance of being able to determine a suitable value of the regularization parameter when the DP cannot be used; see, e.g., [3, 7, 22]. These methods include the L-curve criterion, and generalized cross validation [3].

These methods are outperformed by more recent criteria based on Steins unbiased risk estimate (SURE) [4, 18]. SURE provides an estimate of the mean squared error (MSE), assuming knowledge of the noise distribution and requiring an accurate estimate of its variance [23].

Recently, the fact that the additive noise is the realization of a white random process and, hence, that the restoration residual image must be uncorrelated, has been used not only as an a-posteriori performance evaluation criterion (see, e.g., [21]), but also as a key idea in the design of new fidelity terms [11, 12, 14]. In particular, by evaluating the resemblance of the residue image to a white noise realization, one can check, to some extents, the quality of the restored image. In [1, 8, 20, 21] the measures of residual spectral whiteness have been exploited for adjusting the regularization parameter and/or the number of iterations of the algorithms for deconvolution problems. Comparisons among several state of the art methods have been documented in [2].

## 3 The Class of CNC Variational Models

The considered class of CNC variational models proposed in [15] relies on a general strategy for constructing non-convex non-separable regularizers starting from any convex regularizer $\mathcal{R} : \mathbb{R}^n \to \mathbb{R}$ of the form

$$\mathcal{R}(x) := \Phi(y)\,, \quad y := G(Lx)\,, \tag{3}$$

with $L \in \mathbb{R}^{r \times n}$, $G\colon \mathbb{R}^r \to \mathbb{R}^s$ a possibly nonlinear vector-valued function with $g_i : \mathbb{R}^r \to \mathbb{R}, i = 1, \ldots, s$, representing its scalar-valued components and $\Phi\colon \mathbb{R}^s \to \mathbb{R}$ a sparsity-promoting penalty function [10, 13, 17]. The CNC model associated with the regularizer $\mathcal{R}$ is as follows

$$x_\lambda^* = \operatorname*{argmin}_{x \in \mathbb{R}^n} \mathcal{J}_B(x;\lambda), \quad \mathcal{J}_B(x;\lambda) := \frac{1}{2}\,\|Ax - b\|_2^2 + \lambda\,\mathcal{R}_B(x), \tag{4}$$

with the parameterized non-convex non-separable regularizer $\mathcal{R}_B$ defined by

$$\mathcal{R}_B(x) := \mathcal{R}(x) - \left(\mathcal{R} \,\square\, \tfrac{1}{2}\,\|B \cdot\|_2^2\right)(x), \tag{5}$$

where $\square$ denotes the infimal convolution operator and $B \in \mathbb{R}^{q \times n}$ is a matrix of parameters.

According to Proposition 8 in [15], a sufficient condition for $\mathcal{J}_B$ to be strongly convex—hence, for the variational model in (4) to admit a unique solution—is that the matrix $B$ satisfies

$$B^\mathsf{T} B \prec (1/\lambda) A^\mathsf{T} A. \tag{6}$$

A simple yet effective strategy for constructing a matrix $B^\mathsf{T} B \in \mathbb{R}^{n \times n}$ satisfying the convexity condition in (6) has been presented in [15]. Since the matrix $A^\mathsf{T} A \in \mathbb{R}^{n \times n}$ is symmetric and positive semidefinite, it admits the eigenvalue decomposition

$$A^\mathsf{T} A = V E V^\mathsf{T}, \quad E, V \in \mathbb{R}^{n \times n}, \quad E = \operatorname{diag}(e_1, \ldots, e_n)\,, \quad V^\mathsf{T} V = V V^\mathsf{T} = I_n\,, \tag{7}$$

with $e_i, i = 1, \ldots, n$, indicating the real non-negative eigenvalues of $A^\mathsf{T} A$. By setting

$$B^\mathsf{T} B = \frac{1}{\lambda}\, V \Gamma E V^\mathsf{T}, \quad \Gamma := \operatorname{diag}(\gamma_1, \ldots, \gamma_n)\,, \quad \gamma_i \in [0, 1)\ \forall i \in \{1, 2, \ldots, n\}\,, \tag{8}$$

then (6) is clearly satisfied. A special case is to set a unique parameter $\gamma = \gamma_i \in [0, 1)\ \forall i$, which corresponds to setting $B = \sqrt{\gamma/\lambda}\, A$.

In the present work, which addresses the specific problem of image restoration, we consider as a first regularization function $\mathcal{R}$ in (3) the popular Total Variation (TV) semi-norm [19]. In this case $\Phi$ is the $\ell_1$ norm function and $L = [D_h^{\mathsf{T}}, D_v^{\mathsf{T}}]^{\mathsf{T}}$, with $D_h, D_v \in \mathbb{R}^{n \times n}$ representing finite difference approximations of the first-order partial derivatives along the horizontal and vertical directions, respectively, then we have:

$$\mathcal{R}(x) = \mathrm{TV}(x) = \|G(Lx)\|_1 = \sum_{i=1}^{n} |g_i(Lx)|, \quad g_i(Lx) = \sqrt{(D_h x)_i^2 + (D_v x)_i^2}. \tag{9}$$

It is well known that TV-based reconstructions favor piecewise-constant solutions, but present staircase effects in the restoration of smooth parts of the images. To avoid this artifact, in the reconstruction of piecewise-affine solutions, a second-order extensions of the TV regularizer can be considered which promotes sparsity of the Hessian Schatten norms instead of the gradient norms. That is, the sum of the Schatten $p$-norms of the Hessian matrices computed at every pixel of the image is minimized [16], where, we recall, the Schatten $p$-norm $\|M\|_{\mathcal{S}_p}$ of a matrix $M \in \mathbb{R}^{z \times z}$ is defined by

$$\|M\|_{\mathcal{S}_p} := \left( \sum_{i=1}^{z} \sigma_i^p(M) \right)^{\frac{1}{p}}, \quad p > 0, \tag{10}$$

with $\sigma_i(M)$ indicating the $i$-th singular value of matrix $M$. Let $L = [D_{hh}^{\mathsf{T}}, D_{vv}^{\mathsf{T}}, D_{hv}^{\mathsf{T}}]^{\mathsf{T}}$ with $D_{hh}, D_{vv}, D_{hv} \in \mathbb{R}^{n \times n}$ representing finite difference approximations of second-order derivatives along horizontal, vertical and mixed horizontal/vertical directions, respectively. Then the Hessian Schatten $p$-norm regularizer is defined by

$$\mathcal{R}(x) = \mathcal{S}_p H(x) = \|G(Lx)\|_1 = \sum_{i=1}^{n} |g_i(Lx)|, \quad g_i(Lx) = \left\| \begin{bmatrix} (D_{hh}x)_i (D_{hv}x)_i \\ (D_{vh}x)_i (D_{vv}x)_i \end{bmatrix} \right\|_{\mathcal{S}_p}. \tag{11}$$

We recall that the Schatten $p$-norm reduces to the nuclear norm when $p = 1$.

## 4 Residual Whiteness

Given a realization $\epsilon := \{\epsilon(i, j) \in \mathbb{R} : (i, j) \in \Omega\}$, $\Omega = \{1, 2, \ldots, n_1\} \times \{1, 2, \ldots, n_2\}$ of a 2D $n_1 \times n_2$ random noise process, that is the series of noise values corrupting the particular observed image according to the deterministic degradation model in (1), the *sample* auto-correlation of $\epsilon$ is a function $a_\epsilon$ mapping all the possible lags $(l, m) \in \Theta = \{-(n_1 - 1), \ldots, n_1 - 1\} \times \{-(n_2 - 1), \ldots, n_2 - 1\}$ into a scalar value given by

$$
\begin{aligned}
a_\epsilon(l,m) &:= \frac{1}{n}\left(\epsilon \star \epsilon\right)_{l,m} = \left(\epsilon * \epsilon'\right)_{l,m} \\
&= \frac{1}{n}\sum_{(i,j)\in\Omega} \epsilon(i,j)\,\epsilon(i+l,j+m)\,, \quad (l,m)\in\Theta\,, \quad n = n_1 n_2\,, \qquad (12)
\end{aligned}
$$

where $\star$ and $*$ denote the 2-D discrete correlation and convolution operators, respectively, and where $\epsilon'(i,j) = \epsilon(-i,-j)$. Clearly, for (12) being defined for all lags $(l,m)\in\Theta$, the noise realization $\epsilon$ must be padded with at least $n_1$ samples in the vertical direction and $n_2$ samples in the horizontal direction. We assume here periodic boundary conditions for $\epsilon$, such that $\star$ and $*$ in (12) denote circular correlation and convolution, respectively. If the noise process $\epsilon$ is white, then it is well known that the auto-correlation $a_\epsilon$ satisfies the following asymptotic property:

$$
\lim_{n\to+\infty} a_\epsilon(l,m) = 0 \quad \forall\,(l,m)\in\Theta_0 = \Theta\setminus\{(0,0)\}. \tag{13}
$$

For noise corruptions affecting images of finite dimensions—namely, $n < +\infty$—we can say that the auto-correlation values for all non-zero lags are small. Some important examples of distributions of additive white noises are the uniform, the Gaussian, the Laplacian and the Cauchy [14].

Clearly the nearest to the uncorrupted image is the restored image $x_\lambda^*$, the closer the residual image $r_\lambda^* = b - A\,x_\lambda^*$ is to the realization $\epsilon$ in (1) of a white noise process.

Our proposal is to seek for the regularization parameter value $\lambda^*$ yielding the whitest restoration residual, which can be formally defined as follows:

$$
\lambda^* \in \arg\min_{\lambda\in\mathbb{R}_+} \left\{ W(\lambda) := \mathcal{W}\left(r_\lambda^*\right)\right\}, \tag{14}
$$

with $\mathcal{W}:\mathbb{R}^n \to \mathbb{R}$ one of the two following residual whiteness measures:

$$
\mathcal{W}_1\left(r_\lambda^*\right) = \frac{\sqrt{\sum_{(l,m)\in\Theta_0}\left(a_{r_\lambda^*}(l,m)\right)^2}}{a_{r_\lambda^*}(0,0)}, \quad \mathcal{W}_2\left(r_\lambda^*\right) = \frac{\max_{(l,m)\in\Theta_0}\left|a_{r_\lambda^*}(l,m)\right|}{a_{r_\lambda^*}(0,0)}. \tag{15}
$$

We notice that, according to definition (12), the term $a_{r_\lambda^*}(0,0)$ represents nothing else than the sample variance of the residual image $r_\lambda^*$.

## 5 The Proposed Algorithmic Framework

The proposed bilevel framework consists of an iterative procedure for computing an approximate solution $x_{\lambda^*}^*$ of the class of CNC models proposed in [15] and defined in (4), or also of the associated purely convex models in (2), with $\mathcal{R}$ any sparsity-promoting convex regularizer of the form in (3) and $\lambda^*$ satisfying the whiteness maximality criterion in (14).

In Algorithm 1 we report the main computational steps of the overall proposed bilevel framework for image restoration.

The algorithm starts with a sufficiently small value of the parameter $\lambda$ yielding a large value of the residual whiteness measure $W$ in (14)–(15); at each iteration $\lambda$ is increased by a multiplicative factor $\theta > 1$ to strengthen the effect of the regularization term. The iterative procedure is terminated as soon as the residual whiteness measure stops decreasing, for a certain $\lambda^*$ value. Such a scheme relies on the assumption that the residual whiteness function $W(\lambda)$ in (14) is monotonically decreasing on the $\lambda$ interval between 0 and the function minimizer $\lambda^*$. This property of the whiteness function $W(\lambda)$ is very hard to be proved theoretically but we verified it empirically and the evidence of such behavior is reported in Sect. 6.

At each (outer) iteration $h$ of Algorithm 1, the restored image $x^{(h)}$ is computed—that is, the corresponding optimization problem is solved—by using the Primal-Dual Forward-Backward (PDFB) algorithm described in [15] for the CNC models and the Alternating Direction Method of Multipliers (ADMM) for the associated purely convex models. We remark that, for any given $\lambda$ value, the considered variational models are strongly convex—hence they admit a unique global minimizer—and the PDFB and ADMM minimization algorithms are guaranteed to converge towards such minimizer.

We adopt for efficiency purposes the so called warm-starting strategy to initialize the algorithm at the next inner optimization step using the estimated values at the previous step.

The considered CNC variational approach requires the design of a matrix $B$ satisfying the convexity condition (6) for the functional $\mathcal{J}_{\mathcal{B}}$. Many such matrices $B$ exist, see [15]. In the following experiments, we set $\Gamma$ to be a two-dimensional dc-notch filter defined by $\Gamma = I - H$ where $H$ is a two-dimensional low-pass filter with a dc-gain of unity and $H \leq I$. In our experiments, we set $\gamma = 0.98$ and $H = H_0^{\mathsf{T}} H_0$ where $H_0$ is the most basic two-dimensional low-pass filter: the moving-average filter with square support.

## 6 Numerical Examples

In this section, we report some experimental results aimed at assessing the effectiveness of the automatic parameter selection procedure illustrated in Sects. 4 and 5 for image restoration by using the CNC variational models recently proposed in [15] and briefly outlined in Sect. 3.

We consider the three test images shown in the first row of Fig. 1: `qrcode` which belongs to the class of piecewise constant images, `roof` which is a piecewise affine image, and the popular photographic `cameraman` image. The test images have been synthetically corrupted by space-invariant Gaussian blur generated by the Matlab command `fspecial`('gaussian',`band`,`sigma`) with parameters (`band`,`sigma`) = (5,1.5), and AWGN of standard deviation $\sigma = 40$, so as to obtain the three degraded images shown in the second row of Fig. 1. The `qrcode` and `roof` images are

**Algorithm 1** Bilevel Framework based on Residual Whiteness

**inputs**: degraded image $b \in \mathbb{R}^n$, blur operator $A \in \mathbb{R}^{n \times n}$
**outputs**: regularization parameter $\lambda^* > 0$, restored image $x^*_{\lambda^*} \in \mathbb{R}^n$
**parameters**: $\lambda_{min} > 0, \theta > 1$

initialization: $h = 0, \ \lambda^{(h)} = \lambda_{min}, \ W^{(h)} = +\infty$

**repeat**

· **update iteration counter and regularization parameter:**

· $h = h + 1, \ \lambda^{(h)} = \theta \, \lambda^{(h-1)}$

· **compute restored image by solving the optimization problem:**

· $x^{(h)} = \arg\min_{x \in \mathbb{R}^n} \mathcal{J}_B(x; \lambda^{(h)}) \quad \left(\text{or } \mathcal{J}(x; \lambda^{(h)})\right)$

· **compute residual whiteness:**

· $r^{(h)} = b - Ax^{(h)}$, and $W^{(h)} = \mathcal{W}(r^{(h)})$, by (15)

**until** $W^{(h)} > W^{(h-1)}$

$\lambda^* = \lambda^{(h-1)}, x^*_{\lambda^*} = x^{(h-1)}$

**Fig. 1** Original (first row) and degraded (second row) test images `qrcode` (left column), `roof` (center column) and `cameraman` (right column)

characterized by very sparse first- and second-order derivatives, respectively, hence the convex TV and Schatten 1-norm regularization terms in (9)–(11) and their non-convex non-separable counterparts defined according to (5) are suitable to get good restorations.

Hence, in the experiments we perform restoration by using the proposed bilevel framework outlined in Algorithm 1 applied to the two purely convex variational models

$$\text{TV}-\ell_2: \; x_\lambda^* = \arg\min_{x \in \mathbb{R}^n} \left\{ \frac{1}{2} \|Ax - b\|_2^2 + \lambda \, \text{TV}(x) \right\}, \tag{16}$$

$$\mathcal{S}_1 H-\ell_2: \; x_\lambda^* = \arg\min_{x \in \mathbb{R}^n} \left\{ \frac{1}{2} \|Ax - b\|_2^2 + \lambda \, \mathcal{S}_1 H(x) \right\}, \tag{17}$$

and the two associated CNC counterparts

$$\text{CNC}-\text{TV}-\ell_2: \; x_\lambda^* = \arg\min_{x \in \mathbb{R}^n} \left\{ \frac{1}{2} \|Ax - b\|_2^2 + \lambda \Big( \text{TV} - \text{TV} \,\square\, \tfrac{1}{2} \|B \cdot\|_2^2 \Big)(x) \right\}, \tag{18}$$

$$\text{CNC}-\mathcal{S}_1 H-\ell_2: \; x_\lambda^* = \arg\min_{x \in \mathbb{R}^n} \left\{ \frac{1}{2} \|Ax - b\|_2^2 + \lambda \Big( \mathcal{S}_1 H - \mathcal{S}_1 H \,\square\, \tfrac{1}{2} \|B \cdot\|_2^2 \Big)(x) \right\}. \tag{19}$$

For all the tests, i.e. for all images and all restoration models, the bilevel framework outlined in Algorithm 1 has been used in order to automatically select the regularization parameter $\lambda^*$ yielding the whitest restoration residual according to both the whiteness measures $\mathcal{W}_1, \mathcal{W}_2$ defined in (15); we denote by $\lambda_1^*, \lambda_2^*$ such two optimal values and by $W_1^* := \mathcal{W}_1(\lambda_1^*)$, $W_2^* := \mathcal{W}_2(\lambda_2^*)$ the associated (minimum) whiteness measure values.

The (inner) iterations of the minimization algorithms used to determine the restored image for any given $\lambda$ value—namely, ADMM for the TV$-\ell_2$ and $\mathcal{S}_1 H-\ell_2$ models, PDFB for the CNC$-$TV$-\ell_2$ and CNC$-\mathcal{S}_1 H-\ell_2$ models—are terminated as soon as two successive iterates satisfy

$$\frac{\left\| x^{(h)} - x^{(h-1)} \right\|_2}{\left\| x^{(h-1)} \right\|_2} < 10^{-5}. \tag{20}$$

The quality of the obtained restorations is evaluated by means of both the Signal-to-Noise Ratio (SNR) and the Structural Similarity Index (SSIM). We indicate by $\text{SNR}_1^*$, $\text{SNR}_2^*$ and $\text{SSIM}_1^*$, $\text{SSIM}_2^*$ the SNR and SSIM values of the restored images associated with the optimal values $\lambda_1^*$, $\lambda_2^*$. In order to quantitatively evaluate the ability of the proposed approach in automatically selecting $\lambda$ values yielding restored images of good quality, we also introduce—and compute for each test—the following quantities:

$$\text{LQ}_j^* = 100 \, \frac{\overline{\text{Q}} - \text{Q}_j^*}{\overline{\text{Q}}}, \qquad \text{Q} \in \{\text{SNR}, \text{SSIM}\}, \quad j \in \{1, 2\}, \tag{21}$$

where $\overline{\text{Q}}$ denotes the maximum value of the quality measure - SNR or SSIM—achievable by letting $\lambda$ vary in its domain. These quantities represent the loss of restoration quality (in percentage) yielded by the proposed automatic selection procedure with respect to the maximum achievable.

In Table 1 we report all the obtained quantitative results, whereas in Figs. 2, 3, 4 and 5 we show some visual and graphical results related to the restoration of the `cameraman` test image by the four variational models considered. In particular, the results obtained by using the whiteness measure $\mathcal{W}_1$ are in Figs. 2 and 3, by whiteness measure $\mathcal{W}_2$ in Figs. 4 and 5. Each column in these figures corresponds to a different restoration model. In the third, fourth and fifth row we report the plots of the SNR and SSIM values of the restored image and the plots of the whiteness measure of the restoration residual as functions of the regularization parameter $\lambda$, respectively. The dashed vertical red lines indicate the "optimal" regularization parameter values, namely those yielding the smallest residual whiteness measures. It is worth noticing that for all reported tests the residual whiteness measure function $W(\lambda)$ with both the choices of $\mathcal{W}$ introduced in (15)—shown in the last row of Figs. 2, 3, 4 and 5—exhibit a monotonically decreasing behavior on the $\lambda$ interval between 0 and the functions minimizer $\lambda^*$.

In the first and second row of Figs. 2, 3, 4 and 5 we show the restored images obtained by using such optimal $\lambda$ values and the associated absolute error images, respectively.

In Table 1 the best results are marked in boldface. The results obtained by hand-tuning $\lambda$ (labeled as $\overline{\text{Q}}$) indicate that, as expected, TV-based models perform better on the piecewise constant images `qrcode` and `cameraman` whereas $\mathcal{S}_1 H$-based models outperform TV-based models on the piecewise affine image `roof`. More precisely, the CNC models perform better than their associated purely convex counterparts. This is due to the stronger sparsity-promoting effect produced by non-convex regularization.

For what regards the optimal residual whiteness measures $\text{W}_1^*$ and $\text{W}_2^*$ reported in the last two columns of Table 1, it is worth observing that the lowest results (in boldface) are obtained in correspondence of the best performing models for each restoration test. This in principle should allow to use the proposed automatic parameter selection strategy in order to automatically select the best regularization term for each problem.

Finally, for any given model, the proposed automatic parameter selection strategy seems to perform very well as indicated by the small values of the quality losses $\text{LQ}_j^*$, $j = 1, 2$, reported in Table 1 and visually supported by the plots in the figures.

Visual inspection and comparison of the restored images are consistent with the results in Table 1.

**Table 1** SNR/SSIM results obtained by restoring the test images `qrcode`, `roof` and `cameraman`

| **qrcode** | | $\overline{Q}$ | $Q_1^*$ | $Q_2^*$ | $LQ_1^*$ | $LQ_2^*$ | $W_1^*$ | $W_2^*$ |
|---|---|---|---|---|---|---|---|---|
| Q = SNR | TV-$\ell_2$ | 12.857 | 12.778 | 12.682 | 0.6 | 1.4 | 0.3982 | 1.7928 |
| | CNC-TV-$\ell_2$ | **14.262** | **13.708** | **13.631** | 3.9 | 4.4 | **0.3947** | **1.6594** |
| | $\mathcal{S}_1 H$-$\ell_2$ | 9.052 | 8.836 | 8.520 | 2.4 | 5.9 | 0.4242 | 4.0721 |
| | CNC-$\mathcal{S}_1 H$-$\ell_2$ | 8.994 | 8.847 | 8.514 | 1.6 | 5.3 | 0.4245 | 4.2746 |
| Q = SSIM | TV-$\ell_2$ | 0.803 | 0.803 | 0.801 | 0.0 | 0.3 | 0.3982 | 1.7928 |
| | CNC-TV-$\ell_2$ | **0.869** | **0.850** | **0.849** | 2.2 | 2.3 | **0.3947** | **1.6594** |
| | $\mathcal{S}_1 H$-$\ell_2$ | 0.565 | 0.565 | 0.561 | 0.0 | 0.6 | 0.4242 | 4.0721 |
| | CNC-$\mathcal{S}_1 H$-$\ell_2$ | 0.558 | 0.557 | 0.556 | 0.0 | 0.4 | 0.4245 | 4.2746 |
| **roof** | | $\overline{Q}$ | $Q_1^*$ | $Q_2^*$ | $LQ_1^*$ | $LQ_2^*$ | $W_1^*$ | $W_2^*$ |
| Q = SNR | TV-$\ell_2$ | 22.927 | 22.599 | 22.697 | 1.4 | 1.0 | 0.5096 | 2.2322 |
| | CNC-TV-$\ell_2$ | 22.658 | 22.658 | 22.526 | 0.0 | 0.6 | 0.5100 | 2.2410 |
| | $\mathcal{S}_1 H$-$\ell_2$ | 39.567 | 39.199 | 39.514 | 0.9 | 0.1 | 0.5025 | 2.1634 |
| | CNC-$\mathcal{S}_1 H$-$\ell_2$ | **42.171** | **41.720** | **41.906** | 1.1 | 0.6 | **0.5002** | **2.1481** |
| Q = SSIM | TV-$\ell_2$ | 0.919 | 0.896 | 0.916 | 2.5 | 0.3 | 0.5096 | 2.2322 |
| | CNC-TV-$\ell_2$ | 0.913 | 0.902 | 0.910 | 1.2 | 0.3 | 0.5100 | 2.2410 |
| | $\mathcal{S}_1 H$-$\ell_2$ | 0.999 | 0.999 | 0.999 | 0.0 | 0.0 | 0.5025 | 2.1634 |
| | CNC-$\mathcal{S}_1 H$-$\ell_2$ | **0.999** | **0.999** | **0.999** | 0.0 | 0.0 | **0.5002** | **2.1481** |
| **cameraman** | | $\overline{Q}$ | $Q_1^*$ | $Q_2^*$ | $LQ_1^*$ | $LQ_2^*$ | $W_1^*$ | $W_2^*$ |
| Q = SNR | TV-$\ell_2$ | 11.009 | 10.712 | 10.808 | 2.7 | 1.8 | 0.3950 | 1.6809 |
| | CNC-TV-$\ell_2$ | **11.125** | **10.780** | **10.974** | 3.1 | 1.4 | **0.3946** | **1.6775** |
| | $\mathcal{S}_1 H$-$\ell_2$ | 10.445 | 10.006 | 9.938 | 4.2 | 4.9 | 0.3994 | 1.9923 |
| | CNC-$\mathcal{S}_1 H$-$\ell_2$ | 10.211 | 9.951 | 9.821 | 2.5 | 3.8 | 0.3992 | 2.0405 |
| Q = SSIM | TV-$\ell_2$ | 0.701 | 0.699 | 0.701 | 0.2 | 0.0 | 0.3950 | 1.6809 |
| | CNC-TV-$\ell_2$ | **0.707** | **0.699** | **0.707** | 1.1 | 0.0 | **0.3946** | **1.6775** |
| | $\mathcal{S}_1 H$-$\ell_2$ | 0.663 | 0.663 | 0.663 | 0.0 | 0.0 | 0.3994 | 1.9923 |
| | CNC-$\mathcal{S}_1 H$-$\ell_2$ | 0.650 | 0.648 | 0.650 | 0.3 | 0.0 | 0.3992 | 2.0405 |

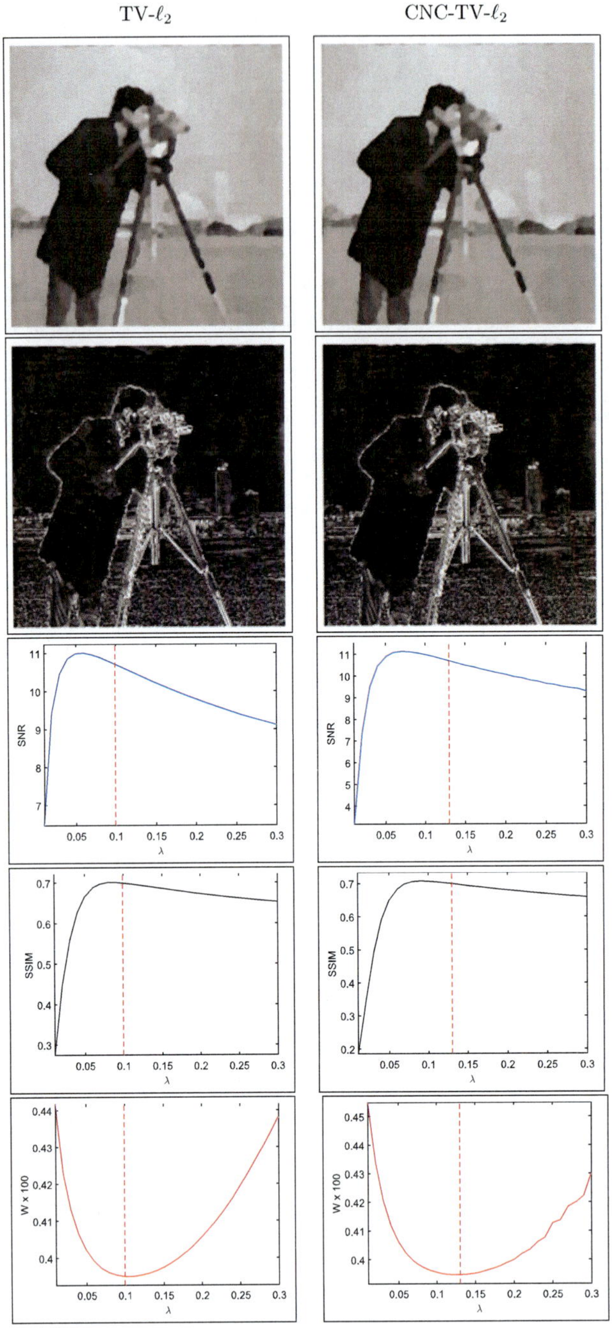


**Fig. 2** Visual/graphical results obtained by using the $\mathcal{W}_1$ residual whiteness measure

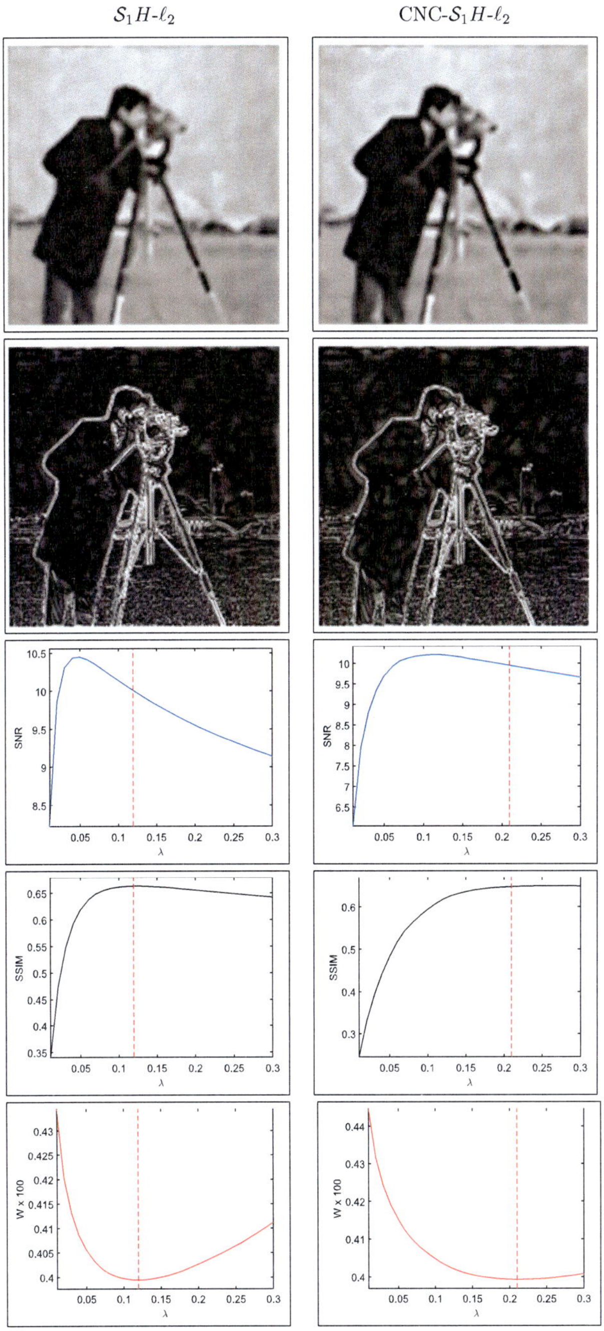


**Fig. 3** Visual/graphical results obtained by using the $\mathcal{W}_1$ residual whiteness measure

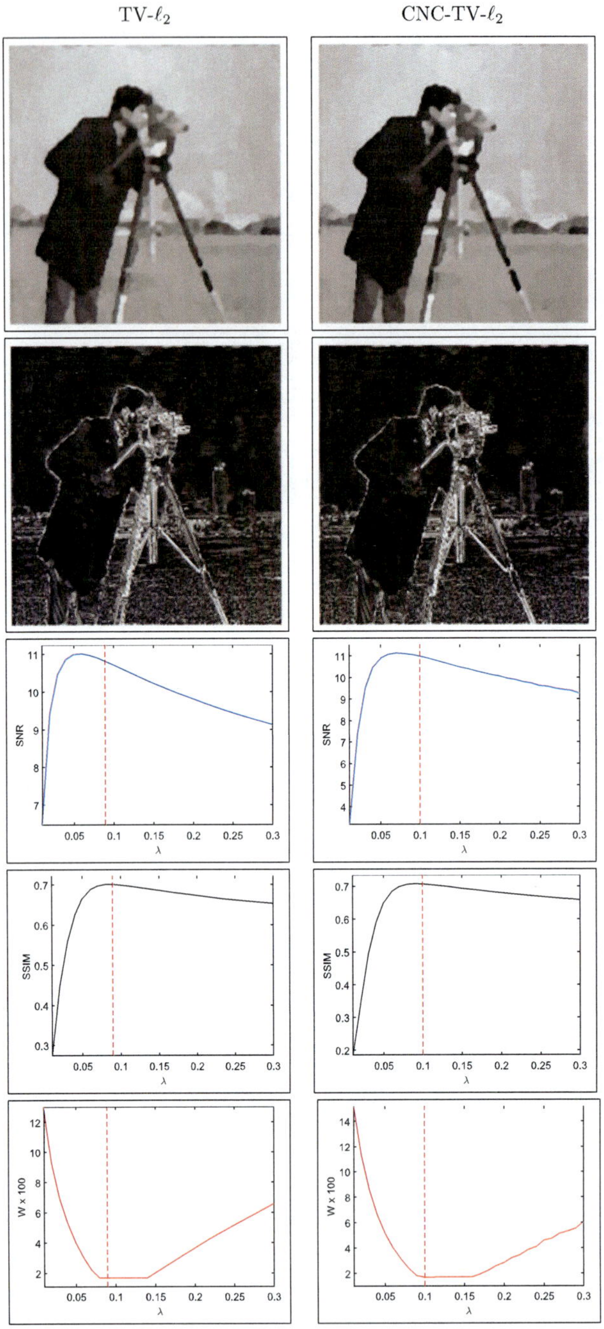


**Fig. 4** Visual/graphical results obtained by using the $\mathcal{W}_2$ residual whiteness measure

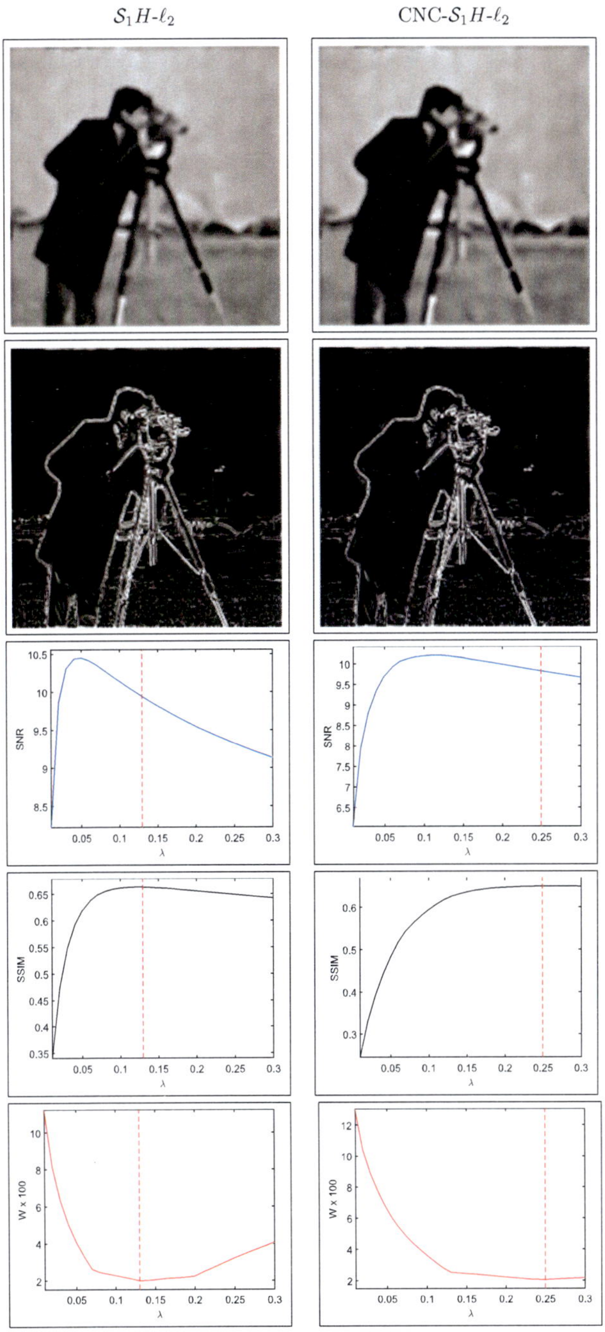


**Fig. 5** Visual/graphical results obtained by using the $\mathcal{W}_2$ residual whiteness measure

## 7 Conclusions

We presented a bilevel framework aimed at equipping the class of CNC variational models for image restoration proposed in [15] with an effective strategy for automatically selecting the regularization parameter based on maximizing the residual whiteness. The idea behind our proposal is that if the recovered image is well estimated, the residual image is spectrally white; on the contrary a poorly restored image exhibits structured artifacts which yield spectrally colored residual images. Numerical results for restoring images characterized by some sparsity properties strongly indicate that the considered class of CNC models with the proposed automatic parameter selection strategy outperforms classical convex models with non-smooth but convex regularizers. The proposed parameter selection strategy makes the considered class of CNC models automatic, in the sense that the regularization parameter is set without requiring any knowledge about the noise variance.

**Acknowledgements** We would like to thank the referees for comments that lead to improvements of the presentation. This research was supported in part by the National Group for Scientific Computation (GNCS-INDAM), Research Projects 2019.

## References

1. M. Almeida, M. Figueiredo, Parameter estimation for blind and non-blind deblurring using residual whiteness measures. IEEE Transactions on Image Processing **22**(7), 2751–2763 (2013)
2. F. Bauer, M.A. Lukas, Comparing parameter choice methods for regularization of ill-posed problems. Mathematics and Computers in Simulation **81**(9), 1795–1841 (2011)
3. D. Calvetti, S. Morigi, L. Reichel, F. Sgallari, Tikhonov regularization and the L-curve for large discrete ill-posed problems. Journal of Computational and Applied Mathematics **123**(1–2), 423–446 (2000)
4. Y.C. Eldar, Generalized SURE for exponential families: Applications to regularization. IEEE Transactions on Signal Processing **57**(2), 471–481 (2009)
5. H.W. Engl, M. Hanke, A. Neubauer, *Regularization of Inverse Problems* (Kluwer, Dordrecht, 1996)
6. U. Haamarik, R. Palm, T. Raus, A family of rules for parameter choice in Tikhonov regularization of ill-posed problems with inexact noise level. Journal of Computational and Applied Mathematics **236**(8), 2146–2157 (2012)
7. P.C. Hansen, D.P. O'Leary, The use of the L-curve in the regularization of discrete ill-posed problems. SIAM Journal on Scientific Computing **14**(6), 1487–1503 (1993)
8. P.C. Hansen, M.E. Kilmer, R.H. Kjeldsen, Exploiting residual information in the parameter choice for discrete ill-posed problems. BIT Numerical Mathematics **46**(1), 41–59 (2006)
9. C. He, C. Hu, W. Zhang, B. Shi, A fast adaptive parameter estimation for total variation image restoration. IEEE Transactions on Image Processing **23**(21), 4954–4967 (2014)
10. A. Lanza, S. Morigi, F. Sgallari, Convex image denoising via non-convex regularization. Lecture Notes in Computer Science (including subseries Lecture Notes in Artificial Intelligence and Lecture Notes in Bioinformatics) **9087**, 666–677 (2015)
11. A. Lanza, S. Morigi, F. Sgallari, A.J. Yezzi, Variational image denoising based on autocorrelation whiteness. SIAM Journal on Imaging Sciences **6**(4), 1931–1955 (2013)
12. A. Lanza, S. Morigi, F. Sgallari, Variational image restoration with constraints on noise whiteness. Journal of Mathematical Imaging and Vision **53**(1), 61–77 (2015)

13. A. Lanza, S. Morigi, F. Sgallari, Convex image denoising via non-convex regularization with parameter selection. Journal of Mathematical Imaging and Vision **56**(2), 195–220 (2016)
14. A. Lanza, S. Morigi, F. Sciacchitano, F. Sgallari, Whiteness constraints in a unified variational framework for image restoration. Journal of Mathematical Imaging and Vision **60**(9), 1503–1526 (2018)
15. A. Lanza, S. Morigi, I.W. Selesnick, F. Sgallari, Sparsity-inducing nonconvex nonseparable regularization for convex image processing. SIAM Journal on Imaging Sciences **12**(2), 1099–1134 (2019)
16. S. Lefkimmiatis, J.P. Ward, M. Unser, Hessian Schatten-norm regularization for linear inverse problems. IEEE Transactions on Image Processing **22**(5), 1873–1888 (2013)
17. A. Parekh, I.W. Selesnick, Convex denoising using non-convex tight frame regularization. IEEE Signal Processing Letters **22**(10), 1786–1790 (2015)
18. S. Ramani, Z. Liu, J. Rosen, J. Nielsen, J.A. Fessler, Regularization parameter selection for nonlinear iterative image restoration and MRI reconstruction using GCV and SURE-based methods. IEEE Transactions on Image Processing **21**(8), 3659–3672 (2012)
19. L.I. Rudin, S. Osher, E. Fatemi, Nonlinear total variation based noise removal algorithms, Physics D, vol. 60(1–4), pp. 259–268, 1992
20. B.W. Rust, D.P. O'Leary, Residual periodograms for choosing regularization parameters for ill-posed problems. Inverse Probl. **24**(3) (2008)
21. B.W. Rust, Parameter selection for constrained solutions to ill-posed problems. Computing Science and Statistics **32**, 333–347 (2000)
22. A.M. Thompson, J.C. Brown, J.W. Kay, D.M. Titterington, A study of methods of choosing the smoothing parameter in image restoration by regularization. IEEE Transactions on Pattern Analysis and Machine Intelligence **13**(4), 326–339 (1991)
23. X. Zhu, P. Milanfar, Automatic parameter selection for denoising algorithms using a no-reference measure of image content. IEEE Transactions on Image Processing **19**(12), 3116–3132 (2010)

# Total Variation Gamma Correction Method for Tone Mapped HDR Images

**Michael K. Ng and Motong Qiao**

**Abstract** Tone mapping methods aim to display a high dynamic range (HDR) image on a common 8-bit liquid crystal display by compressing its dynamic range. Both color rendering and contrast are two important issues in the development of tone mapping methods. In this paper, we propose a variational method to generate low dynamic range (LDR) images by using localized Gamma correction for HDR images to deal with color rendering and contrast issues. Our idea is to employ a weight map that controls localized Gamma correction in each pixel, and the weights are determined by minimizing the differences between the contrast of the original HRD image and that of the LDR image at nearby pixels. By imposing the regularization of the weight map, the total variational term for the weights is incorporated in the objective function for Gamma correction process. Numerical results based on widely-used HDR images are reported to illustrate the effectiveness of the proposed method and the visibility of the details in tone mapped images compared with the other testing methods.

**Keywords** Gamma correction · Total variation · Tone mapping · Dynamic range

## 1 Introduction

In this paper, we study how to convert a high dynamic range (HDR) image to a low dynamic range (LDR) image (or a tone mapped image) such tht a common LCD device can be used to display appropriately. This is called a tone mapping problem in the literature [1]. Contrast adjustment and brightness preservation are the key issues to be addressed in the tone mapping problem. Contrast adjustment refers to

---

Research supported in part by the HKRGC GRF 12200317, 12300218 and 12300519, 17201020 and HKU Grant 104005583.

---

M. K. Ng (✉) · M. Qiao
Department of Mathematics, The University of Hong Kong, Pokfulam Road, Hong Kong, China
e-mail: mng@maths.hku.hk



X.-C. Tai et al. (eds.), *Mathematical Methods in Image Processing and Inverse Problems*,
Springer Proceedings in Mathematics & Statistics 360,
https://doi.org/10.1007/978-981-16-2701-9_7

the compression of the dynamic range of luminance values while preserving the visibility of fine details in tone mapped images. Brightness preservation means that the visual perception of the brightness of tone mapped images should be close to that of original HDR images.

Tone mapping methods are usually employed to generate tone mapped images from HDR images. There are two approaches of tone mapped operators: global methods and local methods. Global tone mapped operators compress pixel values of HDR images by using a uniform scaling function regardless of pixel locations. The main advantage is that the involved computational task is simple and the speed of this method is very fast. It is clear that fine image details may be degraded because of adjusting very low and/or high pixel intensity values. In contrast, local tone mapping methods consider image details and their spatial locations to design a nonuniform scaling function for pixel values compression.

Given an HDR image recording the radiance values of three colors (red, green and blue), the $(i, j)$-th pixel luminance value $H(i, j)$ of an HDR image in RGB color space [2] is given by

$$H(i, j) = 0.2126 R_h(i, j) + 0.7152 G_h(i, j) + 0.0722 B_h(i, j), \tag{1}$$

where $R_h(i, j)$, $G_h(i, j)$ and $B_h(i, j)$ are radiance values of an HDR image at the $(i, j)$-th pixel in RGB color space. The calculation of luminance in other color spaces can be found in [3]. In [4], Tumblin and Rushmeier employed Stevens power-law observer model [5] and developed a global tone mapping method by matching the display device brightness and the real world brightness. Their method is to apply Gamma correction to the HDR luminance to obtain the LDR luminance:

$$L(i, j) = c_1 H(i, j)^{\alpha}. \tag{2}$$

Here $\alpha$ is a positive number determined by the real world luminance and the display output luminance, and $c_1$ is a positive number to control the scale of the output. In [6], Ward simplified the above formula by using a linear scaling formula: $L(i, j) = c_2 H(i, j)$. Ward proposed to compute the scaling number $c_2$ based on the minimum luminance difference between the display and the real world scene detected by eye. In [7], Schlick proposed a rational scaling formula to obtain the tone mapped luminance:

$$L(i, j) = \frac{c_3 H(i, j)}{c_4 H(i, j) + \max_{i,j}\{H(i, j)\}}, \tag{3}$$

where $c_3$ and $c_4$ are positive numbers set by the user based on the smallest intensity level on the display device.

Local tone mapping methods utilize pixel information to determine different scaling functions for HDR luminance values. Chiu et al. [8] considered using a nonuniform mapping function based on the reciprocal of the local mean of HDR luminance values $S(i, j)$ and the LDR luminance is given by

$$L(i, j) = \frac{H(i, j)}{S(i, j)}. \tag{4}$$

The local mean $S(\cdot, \cdot)$ is obtained based on the low pass filter on the HDR luminance $H(\cdot, \cdot)$. This approach preserves the visibility of fine details quite well. However, halo effects may appear at the transition between dark regions and bright regions due to the gradient inversion. In [9], Ashikhmin developed a nonuniform adaptation to compute the tone mapped luminance:

$$L(i, j) = c_5 \frac{\phi(H(i, j)) - \phi(H_{\min})}{\phi(H_{\max}) - \phi(H_{\min})}, \tag{5}$$

where $c_5$ is the maximum luminance value a display output can be generated, and $\phi$ is a piecewise capacity function in which different luminance domains have different responses to handle dark and bright regions in the picture, and $H_{\max} = \max_{(i,j)}\{H(i, j)\}$ and $H_{\min} = \min_{(i,j)}\{H(i, j)\}$. However, experimental results in [9] have shown that artifacts appear at the regions near the edges.

In [10], Tumblin studied the problem of color rendering and proposed to apply the Gamma correction to each color channel scaled by the HDR luminance. The pixel values of red, green and blue channels are further weighted by the logarithm of the HDR luminance ($K(i, j) = \log H(i, j)$). The resulting scheme is given as follows:

$$\text{Red channel: } R_l(i, j) = \left[\frac{R(i, j)}{H(i, j)}\right]^{\gamma} \times K(i, j) \tag{6}$$

$$\text{Green channel: } G_l(i, j) = \left[\frac{G(i, j)}{H(i, j)}\right]^{\gamma} \times K(i, j) \tag{7}$$

$$\text{Blue channel: } B_l(i, j) = \left[\frac{B(i, j)}{H(i, j)}\right]^{\gamma} \times K(i, j) \tag{8}$$

where $\gamma$ is a positive number. In [11], Durand and Dorsey applied the bilateral filtering technique [12] to differentiate edges and smooth regions for tone mapped luminance. The output of bilateral filter contains the decomposition of $H(\cdot, \cdot)$ into two layers: a base layer (i.e., smooth regions) $B(\cdot, \cdot)$ and a detail layer (i.e., edges). The next step is to employ $B(\cdot, \cdot)$ in (2) instead of $H(\cdot, \cdot)$ and combine such compressed the base layer and the detail layer together to obtain the tone mapped luminance. Experimental results have shown that the bilateral-based tone mapping method is effective. In order to implement bilateral filtering, four parameters are required to be set in Gaussian kernels used in the filtering. Recently, Choudhury [13] extended this idea to employ the trilateral filter in which only one parameter is required.

Similar to the bilateral-based tone mapping method, the retinex-based tone-mapping method [14] is studied and developed based on the retinex theory [15]. The idea is to use the retinex algorithm to decompose HDR luminance into two parts: illuminance (base layer) and reflectance (detail layer), and then generate tone mapped luminance. In [14], Drago et al. conducted several psychophysical experiments to determine the required parameters in the retinex algorithm so that the most "natural" looking of tone mapped images can be created. In [16], Meylan proposed to use the first principal component of the linearly-encoded luminance image as an input, and applied a power function to generate an initial global tone mapped image. Then a local adaptation is applied to the global tone mapped image by using a surround-based retinex method to obtain the resulting tone mapped image. Recently, Kim [17] investigated a new $k$-factor decision method ($k$-factor is one of the parameters in retinex algorithm) to enhance the appearance and naturalness of tone mapped images in the compression process. It is interesting to note that these retinex-based tone mapping methods are aimed at achieving a perceptually natural scene of tone mapped images to the observers.

Another approach of local tone mapping methods is to use histogram adjustment techniques on the cumulative distribution of $H(\cdot,\cdot)$ to figure out how to map HDR values to LDR values [18]. Here the crucial issue is to select the number of histogram bins and the width of each bin. In [19], Duan et al. addressed the problem of bin width determination by using histogram equalization techniques. Their idea is to divide the dynamic range of HDR luminance into many non-overlapping sub-ranges of LDR luminance recursively where the histogram in each sub-range should be equalized by finding a suitable bin width. The computational cost of this method is quite costly. In [20], Qiu and Duan formulated the tone mapping problem as a minimization problem as follows:

$$\min_{s_1,s_2,\cdots,s_{255}} \sum_{k=1}^{255}\left(s_k-\frac{k}{256}\right)^2+\lambda\sum_{k=1}^{255}\left(\int_0^{s_k} h(x)dx-\frac{k}{256}\right)^2 \tag{9}$$

where the display levels of LDR luminance are the integers in between 0 and 255, the variables $s_k$ are the corresponding end-points of the bins in the HDR luminance, $h(x)$ is the histogram of the HDR image and $\lambda$ is a parameter to control the balance between the two terms. The first term is used to construct a uniform partition of end-points and the second term is used to determine end-points based on histogram equalization. However, there is no closed form solution for (9). In [21], Qiu et al. further reformulated (9). Their idea is to find the number of pixels in each bin instead of the end-points of the bins so that a closed form solution of the new optimization problem can be obtained. Experimental results have shown the performance of this approach is pretty well.

In [22], Shan et al. proposed to minimize the difference between the LDR and HRD images. They assumed that each pixel has a linear relation with its local neighboring pixels. Their energy minimization problem is given as follows:

$$\min_{\{L(i,j)\},\{p_k\},\{q_k\}} \sum_{(i,j)} \sum_{(u,v)\in\mathcal{N}(i,j)} [L(i,j)-p_kH(u,v)-q_k]^2 + \lambda(p_k-t_k)^2, \tag{10}$$

where $p_k$ is a pixel-wise parameter mainly responsible for scaling the luminance value $H(i, j)$, $q_k$ accounts for intensity offset adjustment, $\mathcal{N}(i, j)$ denotes the neighborhood region centered at the $(i, j)$-th pixel, $t_k$ is a positive constant to guide the value of $p_k$, and $\lambda$ is a positive parameter to balance between the two terms. Given an initial $L(i, j)$, the minimum value for $p_k$ and $q_k$ can be calculated. Then Shan et al. minimized the energy functional in (10) with respect to $L(i, j)$ where $p_k$ and $q_k$ are fixed. This minimization process alternates between $L(i, j)$ and $\{p_k, q_k\}$ until the iterates converge, and the optimized tone mapped luminance can be obtained.

### *1.1 The Contribution*

In this paper, we consider both Gamma correction and total variation based methods to deal with color rendering and contrast issues in the tone mapping problem. Our idea is to design a weight map such that a Gamma correction can be applied to local HDR luminance. Here the weights are determined by minimizing the differences between the contrast of the original HDR image and the contrast of the tone mapped image at neighbor pixels. The advantage of this approach is to preserve local contrast and keep the detailed information of an HDR image into an LDR image. On the other hand, we impose the regularization among the weights, the total variation regularization of the weight map is incorporated in the objective function. The resulting optimization model consisting of the data-fitting term and the regularization term is convex, and it can be solved by using many fast convex optimization solvers. In particular, the alternating direction of multiplier method is used to test several widely-used HDR images. The LDR results by the proposed model are compared with the other testing methods. It can be shown that the proposed model can provide visually very good LDR images.

The outline of this paper is given as follows. In Sect. 2, we present the proposed model and the algorithm is given. In Sect. 3, numerical examples are presented to demonstrate the effectiveness of the proposed model and the algorithm. Finally, some concluding remarks are given in Sect. 4.

## 2 The Proposed Model

Both color rendering and contrast are two important issues in the generation of LDR images from HDR images. In [23], it has been studied that human perception is more sensitive to colorfulness at high luminance levels than at low luminance levels. Also we are more sensitive to contrast at low luminance levels than at high luminance levels. In our proposal, we employ a nonuniform weight $w(i, j)$ in the color rendering and contrast model such that these two components can be adapted to the local information of the given HDR image. More precisely, the LDR values of the red, green and blue channels are given by

$$\text{Red channel: } R_l(i,j) = \left[\frac{R(i,j)}{H(i,j)}\right]^{a/w(i,j)} \times K(i,j)^{bw(i,j)} \tag{11}$$

$$\text{Green channel: } G_l(i,j) = \left[\frac{G(i,j)}{H(i,j)}\right]^{a/w(i,j)} \times K(i,j)^{bw(i,j)} \tag{12}$$

$$\text{Blue channel: } B_l(i,j) = \left[\frac{B(i,j)}{H(i,j)}\right]^{a/w(i,j)} \times K(i,j)^{bw(i,j)}, \tag{13}$$

where $a$ and $b$ are two positive numbers to control the scaling of the color rendering and contrast components in the Gamma correction process. We note that the color rendering component in (11) is similar to that in (6) and the contrast component in (11) is similar to that in (2). However, a nonuniform map is used in the model. At the low (high) luminance level, the contrast component $K(i,j)^{bw(i,j)}$ should be processed by making $w(i,j)$ to be large (small). Similarly, at the high (low) luminance level, the color saturation terms $[R(i,j)/H(i,j)]^{a/w(i,j)}$, $[G(i,j)/H(i,j)]^{a/w(i,j)}$ and $[B(i,j)/H(i,j)]^{a/w(i,j)}$ should be processed making $1/w(i,j)$ to be large (small).

The ratio between the $(i,j)$-th pixel of the HDR image and its neighborhood $(u,v)$-th pixel of the HDR image:

$$\log\frac{R(i,j)}{R(u,v)}, \quad \log\frac{G(i,j)}{G(u,v)}, \quad \log\frac{B(i,j)}{B(u,v)},$$

can be viewed as the contrast of the HDR image at $i$-th pixel. In order to control localized Gamma correction in (11), the weights are determined by minimizing the difference between the contrast of the HDR image and the contrast of the LDR image:

$$\log\frac{R(i,j)}{R(u,v)} - \log\frac{R_l(i,j)}{R_l(u,v)}, \quad \log\frac{G(i,j)}{G(u,v)} - \log\frac{G_l(i,j)}{G_l(u,v)},$$

$$\log\frac{B(i,j)}{B(u,v)} - \log\frac{B_l(i,j)}{B_l(u,v)}.$$

The summarized data-fitting term for the red, green an blue channels is given by

$$\Phi(\mathbf{W}) \equiv \sum_{(u,v)\in\mathcal{N}(i,j)} \left( \left\| \log\frac{R(i,j)}{R(u,v)} - \log\frac{R_l(i,j)}{R_l(u,v)} \right\|_2^2 + \left\| \log\frac{G(i,j)}{G(u,v)} - \log\frac{G_l(i,j)}{G_l(u,v)} \right\|_2^2 + \left\| \log\frac{B(i,j)}{B(u,v)} - \log\frac{B_l(i,j)}{B_l(u,v)} \right\|_2^2 \right), \tag{14}$$

where $\mathbf{W} = [w_{i,j}]$ is a vector containing the unknown weights at all the pixel locations, and $\mathcal{N}(i, j)$ is the neighborhood pixels with respect to the $(i, j)$-th pixel location. For instance, the pixels in the three-by-three window centered at the $(i, j)$-th pixel is employed in the experimental section. We note that

$$\log \frac{R(i, j)}{R(u, v)} - \log \frac{R_l(i, j)}{R_l(u, v)} = \log \frac{R(i, j)}{R_l(i, j)} - \log \frac{R(u, v)}{R_l(u, v)}.$$

Therefore, we observe that the data-fitting term can be considered to calculate the the difference between the original HRD image and the LDR image in the logarithm domain and force the difference at nearby pixels to be about the same.

In order to minimize the differences among the weights at the nearby pixel locations, the total variation regularization $TV(\mathbf{W})$ of the weight map $\mathbf{W}$ is incorporated in the objective function. The resulting objective function is given as follows:

$$\min_{\mathbf{W}} TV(\mathbf{W}) + \frac{\mu}{2}\Phi(\mathbf{W}) \tag{15}$$

where $\mu$ is a positive number to control the balance between the data-fitting term and the regularization term.

## 2.1 The Algorithm

In this subsection, we develop an algorithm to solve (15). We first rewrite the data-fitting term. For the red channel, we note by (11) that

$$\begin{aligned}
&\log \frac{R(i, j)}{R(u, v)} - \log \frac{R_l(i, j)}{R_l(u, v)} \\
&= \log \frac{R(i, j)}{R(u, v)} - \log \left( \frac{\left[\frac{R(i, j)}{H(i, j)}\right]^{-bw(i,j)} \times K(i, j)^{aw(i,j)}}{\left[\frac{R(u, v)}{H(u, v)}\right]^{-bw(u,v)} \times K(u, v)^{aw(u,v)}} \right) \\
&= \log \frac{R(i, j)}{R(u, v)} + w(i, j)\left[a \log \frac{R(i, j)}{H(i, j)} - b \log K(i, j)\right] - \\
&\quad w(u, v)\left[a \log \frac{R(u, v)}{H(u, v)} - b \log K(u, v)\right] \\
&= R_c(i, j) + w(i, j)R_s(i, j) - w(u, v)R_s(u, v),
\end{aligned} \tag{16}$$

where

$$R_c(i, j, u, v) = \log \frac{R(i, j)}{R(u, v)}$$

and

$$R_s(i, j) = a \log \frac{R(i, j)}{H(i, j)} - b \log K(i, j).$$

Similarly, we have

$$\log \frac{G(i, j)}{G_l(i, j)} - \log \frac{G(u, v)}{G_l(u, v)}$$
$$= G_c(i, j, u, v) + w(i, j)G_s(i, j) - w(u, v)G_s(u, v) \quad (17)$$

and

$$\log \frac{B(i, j)}{B_l(i, j)} - \log \frac{B(u, v)}{B_l(u, v)}$$
$$= B_c(i, j, u, v) + w(i, j)B_s(i, j) - w(u, v)B_s(u, v) \quad (18)$$

where

$$G_c(i, j, u, v) = \log \frac{G(i, j)}{G(u, v)}, \quad B_c(i, j, u, v) = \log \frac{B(i, j)}{B(u, v)},$$

$$G_s(i, j) = a \log \frac{G(i, j)}{H(i, j)} - b \log K(i, j),$$

and

$$B_s(i, j) = a \log \frac{B(i, j)}{H(i, j)} - b \log K(i, j).$$

For an $n$-by-$m$ given HDR image, we generate an $n$-by-$m$ LDR image. According to (16), (17) and (18), we have the stencil values for the $(i, j)$-th pixel and its neighborhood pixels described in Tables 1, 2, 3 and 4 in the data-fitting term. By using the lexicographical ordering, we form

$$\mathbf{W} = [w(1, 1), \cdots, w(1, m), w(2, 1), \cdots, w(n, m)]^t,$$

$$\mathbf{R}_c = [R_c(1, 1), \cdots, R_c(1, m), R_c(2, 1), \cdots, R_c(n, m)]^t,$$

$$\mathbf{G}_c = [G_c(1, 1), \cdots, G_c(1, m), G_c(2, 1), \cdots, G_c(n, m)]^t,$$

$$\mathbf{B}_c = [B_c(1, 1), \cdots, B_c(1, m), B_c(2, 1), \cdots, B_c(n, m)]^t,$$

and **R**, **G** and **B** are the $n$-by-$n$ block matrix with $m$-by-$m$ matrix block for the red, green and blue channels respectively:

$$[\mathbf{R}]_{k,k} = R_s(i, j), \quad [\mathbf{G}]_{k,k} = G_s(i, j),$$

$$[\mathbf{B}]_{k,k} = B_s(i, j),$$

**Table 1** The pixel $(i, j)$-th location and its neighborhood pixels for $1 \le i \le n$ and $1 \le j \le m$

| | | | | |
|---|---|---|---|---|
| $\ddots$ | $\vdots$ | $\vdots$ | $\vdots$ | $\iddots$ |
| $\cdots$ | $(i-1, j+1)$ | $(i, j+1)$ | $(i+1, j+1)$ | $\cdots$ |
| $\cdots$ | $(i-1, j)$ | $(i, j)$ | $(i+1, j)$ | $\cdots$ |
| $\cdots$ | $(i-1, j-1)$ | $(i, j-1)$ | $(i+1, j-1)$ | $\cdots$ |
| $\iddots$ | $\vdots$ | $\vdots$ | $\vdots$ | $\ddots$ |

**Table 2** The red channel stencil values at the corresponding the pixel locations in Table 1

| | | | | |
|---|---|---|---|---|
| $\ddots$ | $\vdots$ | $\vdots$ | $\vdots$ | $\iddots$ |
| $\cdots$ | $-R_s(i-1, j+1)$ | $-R_s(i, j+1)$ | $-R_s(i+1, j+1)$ | $\cdots$ |
| $\cdots$ | $-R_s(i-1, j)$ | $R_s(i, j)$ | $-R_s(i+1, j)$ | $\cdots$ |
| $\cdots$ | $-R_s(i-1, j-1)$ | $-R_s(i, j-1)$ | $-R_s(i+1, j-1)$ | $\cdots$ |
| $\iddots$ | $\vdots$ | $\vdots$ | $\vdots$ | $\ddots$ |

**Table 3** The green channel stencil values at the corresponding the pixel locations in Table 1

| | | | | |
|---|---|---|---|---|
| $\ddots$ | $\vdots$ | $\vdots$ | $\vdots$ | $\iddots$ |
| $\cdots$ | $-G_s(i-1, j+1)$ | $-G_s(i, j+1)$ | $-G_s(i+1, j+1)$ | $\cdots$ |
| $\cdots$ | $-G_s(i-1, j)$ | $G_s(i, j)$ | $-G_s(i+1, j)$ | $\cdots$ |
| $\cdots$ | $-G_s(i-1, j-1)$ | $-G_s(i, j-1)$ | $-G_s(i+1, j-1)$ | $\cdots$ |
| $\iddots$ | $\vdots$ | $\vdots$ | $\vdots$ | $\ddots$ |

**Table 4** The blue channel stencil values at the corresponding the pixel locations in Table 1

| | | | | |
|---|---|---|---|---|
| $\ddots$ | $\vdots$ | $\vdots$ | $\vdots$ | $\iddots$ |
| $\cdots$ | $-B_s(i-1, j+1)$ | $-B_s(i, j+1)$ | $-B_s(i+1, j+1)$ | $\cdots$ |
| $\cdots$ | $-B_s(i-1, j)$ | $B_s(i, j)$ | $-B_s(i+1, j)$ | $\cdots$ |
| $\cdots$ | $-B_s(i-1, j-1)$ | $-B_s(i, j-1)$ | $-B_s(i+1, j-1)$ | $\cdots$ |
| $\iddots$ | $\vdots$ | $\vdots$ | $\vdots$ | $\ddots$ |

for $1 \le k \le nm$ with $k = (i-1) \times m + j$; and

$$[\mathbf{R}]_{k,l} = -R_s(u, j), \quad [\mathbf{G}]_{k,l} = -G_s(u, j),$$

$$[\mathbf{B}]_{k,l} = -B_s(u, j),$$

for $1 \leq k \neq l \leq nm$ with $k = (i-1) \times m + j$ and $l = (u-1) \times m + v$. Therefore, the data-fitting term is given by

$$\Phi(\mathbf{W}) = \|\mathbf{RW} - \mathbf{R}_c\|_2^2 + \|\mathbf{GW} - \mathbf{G}_c\|_2^2 + \|\mathbf{BW} - \mathbf{B}_c\|_2^2.$$

The regularization term is given by

$$\left\| \begin{bmatrix} \mathbf{D} \otimes \mathbf{I} \\ \mathbf{I} \otimes \mathbf{D} \end{bmatrix} \mathbf{W} \right\|_2 = \|\mathbf{D}_1 \mathbf{W}\|_2$$

where $\mathbf{D}$ is the first-order finite difference matrix. The objective function in (15) can be re-written as follows:

$$\min_{\mathbf{W}} \|\mathbf{D}_1 \mathbf{W}\|_2 + \frac{\mu}{2} \left( \|\mathbf{RW} - \mathbf{R}_c\|_2^2 + \|\mathbf{GW} - \mathbf{G}_c\|_2^2 + \|\mathbf{BW} - \mathbf{B}_c\|_2^2 \right) \tag{19}$$

The objective function in (19) is convex, and the minimization problem can be solved efficiently by many convex optimization solvers. Here we present the alternating method of multipliers to solve the problem. The main idea is use an auxiliary variable $\mathbf{Y}$ to equal to $\mathbf{D}_1\mathbf{W}$, and set up the the augmented Lagrangian equation of (19):

$$\begin{aligned} &\mathcal{L}(\mathbf{W}, \mathbf{Y}, \Lambda) \\ &= \|\mathbf{Y}\|_2 + \Lambda^t(\mathbf{Y} - \mathbf{D}_1\mathbf{W}) + \frac{\beta}{2}\|\mathbf{Y} - \mathbf{D}_1\mathbf{W}\|_2^2 + \\ &\frac{\mu}{2} \left( \|\mathbf{RW} - \mathbf{R}_c\|_2^2 + \|\mathbf{GW} - \mathbf{G}_c\|_2^2 + \|\mathbf{BW} - \mathbf{B}_c\|_2^2 \right), \end{aligned} \tag{20}$$

where $\Lambda$ is the Lagrangian multiplier and $\beta$ is a positive number to force that the linear constraint $\mathbf{Y} = \mathbf{D}_1\mathbf{W}$ is satisfied. The iterative algorithm of the alternating direction of multipliers is given as follows:

**Step 1:** Initialize $\mathbf{W}^0$, $\Lambda^0$ and set $k = 0$

**Step 2:** Fix $\mathbf{W}^k$, update $\mathbf{Y}^{k+1}$ by:

$$\min_{\mathbf{Y}} \|\mathbf{Y}\|_2 + \frac{\beta}{2} \left\| \mathbf{Y} - (\mathbf{D}_1\mathbf{W}^k + \frac{\Lambda^k}{\beta}) \right\|_2^2$$

The above minimization subproblem can be easily solved by using the shrinkage procedure:

$$\mathbf{Y}^{k+1} = \max\left\{\left\|\mathbf{D}_1\mathbf{W}^k + \frac{\Lambda^k}{\beta}\right\|_2 - \frac{1}{\beta}, 0\right\} \times \frac{\mathbf{D}_1\mathbf{W}^k + \Lambda^k/\beta}{\|\mathbf{D}_1\mathbf{W}^k + \Lambda^k/\beta\|_2} \tag{21}$$

**Step 3:** Fix $\mathbf{Y}^{k+1}$, update $\mathbf{W}^{k+1}$ by:

$$\min_{\mathbf{W}} \frac{\beta}{2}\left\|\mathbf{Y} - (\mathbf{D}_1\mathbf{W} + \frac{\Lambda^k}{\beta})\right\|_2^2 + \frac{\mu}{2}\left(\|\mathbf{RW} - \mathbf{R}_c\|_2^2 + \|\mathbf{GW} - \mathbf{G}_c\|_2^2 + \|\mathbf{BW} - \mathbf{B}_c\|_2^2\right)$$

The above minimization subproblem is quadratic in $\mathbf{W}$, and the corresponding normal equation is solved:

$$\left[\mathbf{D}_1^t\mathbf{D}_1 + \frac{\mu}{\beta}\left(\mathbf{R}^t\mathbf{R} + \mathbf{G}^t\mathbf{G} + \mathbf{B}^t\mathbf{B}\right)\right] = \mathbf{D}_1^t\left(\mathbf{Y} - \frac{\Lambda}{\beta}\right) + \frac{\mu}{\beta}\left(\mathbf{R}^t\mathbf{R}_c + \mathbf{G}^t\mathbf{G}_c + \mathbf{B}^t\mathbf{B}_c\right). \tag{22}$$

The stencil values of $\mathbf{D}_1^t\mathbf{D}_1$ is given in Table 5. Then $\mathbf{W}$ can be calculated by solving the above linear equation.

**Step 4:** Fix $\mathbf{W}^{k+1}$, $\mathbf{Y}^{k+1}$, update $\Lambda^{k+1}$ by

$$\Lambda^{k+1} = \Lambda^k + \beta(\mathbf{Y}^{k+1} - \mathbf{D}_1\mathbf{W}^{k+1}) \tag{23}$$

**Step 5:** Iterate Steps 2, 3 and 4 until $\|\mathbf{W}^{k+1} - \mathbf{W}^k\|_2 \leq \epsilon$.

In the above algorithm, the computation of Step 2 is to perform entry-wise calculation in (21) and its computational cost is of $O(nm)$ operations. It is required to solve a linear system in (22) in Step 3. When $\mathcal{N}(i, j)$ involves the $k$-by-$k$ window centered at the $(i, j)$-th pixel location, the matrices $\mathbf{R}$, $\mathbf{G}$ and $\mathbf{B}$ are sparse, and each

**Table 5** The stencil values of $\mathbf{D}_1^t\mathbf{D}_1$ at the corresponding the pixel locations in Table 1

| $\ddots$ | $\vdots$ | $\vdots$ | $\vdots$ | $\iddots$ |
|---|---|---|---|---|
| ... | 0 | −1 | 0 | ... |
| ... | −1 | 4 | −1 | ... |
| ... | 0 | −1 | 0 | ... |
| $\iddots$ | $\vdots$ | $\vdots$ | $\vdots$ | $\ddots$ |

row of these matrices has $k^2$ nonzero entries. For the matrix $\mathbf{D}_1'\mathbf{D}_1$, each row has five entries. We can employ an iterative method to solving the linear system in (22). The computational cost per iteration is of $O(k^2nm)$ operations. In [24], an inexact alternating direction method of multipliers can be employed in the framework. The computational cost of the proposed method is of $O(k^2nm)$ operations. In the next section, we show experimental results to demonstrate the usefulness of the proposed model and algorithm.

## 3 Experimental Results

In this section, we test the proposed tone mapping method on several widely-tested HDR images. The following parameters are set for all the testing images as follows: $\mathcal{N}(i, j)$ is a 3-by-3 window centered at the $(i, j)$-th pixel location; the stopping criterion of $\epsilon$ is $1 \times 10^{-3}$; the initial guess of $\mathbf{W}^0$ is a vector of all ones; the initial Lagrangian multipliers $\Lambda$ is a zero vector; the penalty parameter $\beta$ is equal to 10; the scaling parameters $a$ and $b$ are 0.6 and 1.2, and the regularization parameter $\mu$ is $10^3$. In the tests, we compared the proposed method with the other tone mapping methods. These methods include Drago's adaptive logarithmic method [25], Ashikhmin's method [9], Banterle's method [26], Durand's bilateral filter based method [11], Fattal's gradient domain based method [27], Reinhard's photographic method [28], Paris's method [29], Yee's segmentation based method [30]. As a comparison, we also give LDR images when the weight map is set to be uniform, i.e., $\mathbf{W}$ is a vector of all ones.

In Figs. 1, 3, 5, 7, 9 and 11, we show the LRD pictures obtained by different methods. In Figs. 2, 4, 6, 8, 10 and 12, we display the zoomed regions of LDR images obtained by different methods in Figs. 1, 3, 5, 7, 9 and 11 so that we can evaluate their visual performance. In general, we find that the nonuniform map determined by the proposed method can provide more visible detailed information of testing images. For all five testing images, it is clear that the visual performance of using nonuniform map is better than that of using uniform map ($\mathbf{W}$ is a vector of all ones). In Fig. 1, the LDR pictures by Fattal's and Ashikhmin's methods show good visibility of details but their contrast are quite low. The LDR pictures by the bilateral filter based methods (Durand's method and Paris's method) show a good contrast, but their color in in the sky and cloud seems bleached. In Fig. 2, we see that the contrast of LDR picture generated by the proposed method is better than those by the other methods. In Figs. 3 and 4, the door entrance are not clear in the LDR pictures by Drago's, Ashikhmin's, Banterle's, Durand's, Reinhard's and Paris's methods. We see that the entrance structure is more clear in the tone mapped image by the proposed method than the pictures by Fattal's and Yee's methods. In Fig. 5, the color rendering of the LDR picture by the proposed method is better than those of the other pictures, especially in the region of the attached lamps (see 6). Also the items attached at handling wall in the right hand side of the LDR pictures by the other methods are not shown very clearly. The proposed method can give these items with a very good

(a) The proposed method with $\mu = 1$

(b) A uniform weight map $\mathbf{W} = \mathbf{1}$

(c) Drago's method [16]

(d) Ashikhmin's method [11]

(e) Banterle's method [28]

(f) Durand's method [12]

(g) Fattal's method [29]

(h) Reinhard's method [30]

(i) Paris's method [31]

(j) Yee's method [32]

**Fig. 1** Vinesunset scene

**Fig. 2** The zoomed region of Vinesunset scene

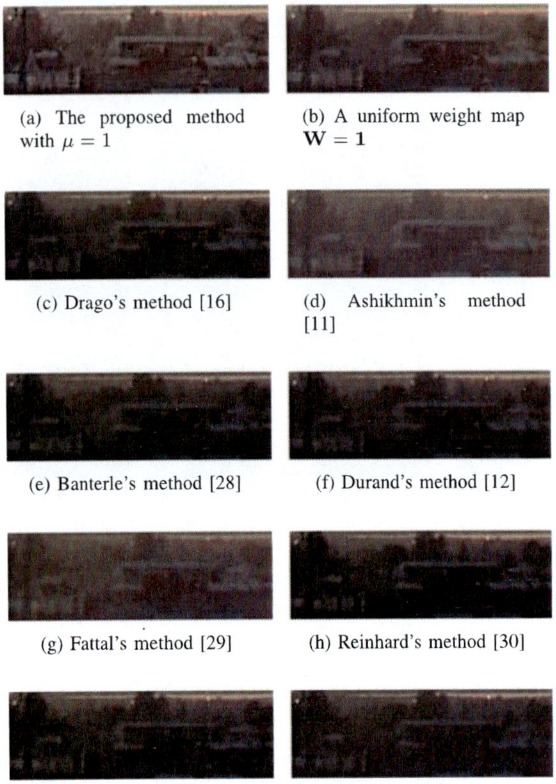

(a) The proposed method with $\mu = 1$

(b) A uniform weight map $\mathbf{W} = \mathbf{1}$

(c) Drago's method [16]

(d) Ashikhmin's method [11]

(e) Banterle's method [28]

(f) Durand's method [12]

(g) Fattal's method [29]

(h) Reinhard's method [30]

(i) Paris's method [31]

(j) Yee's method [32]

visual appearance. In Fig. 7, the proposed method gives a LDR picture with both good contrast and color rendering. In particular, the windows shown in the zoomed region in 8 look very good visually compared with those by the other methods. Their lightening and colors are not natural in the LDR images by the other methods. In Figs. 9 and 10, the fine details such as car plate number by the proposed method can be seen clearly. In Figs. 11 and 12, the contrast of the LDR image by the proposed method is very good.

In general, we find that the proposed method can provide the details in both bright and dark regions of the scene which has higher visibility than those by the other tone mapping methods. The main reason is that the data-fitting term is designed by using (i) both Gamma correction in color rendering and luminance terms, and (ii) the nonuniform weight map such that different weights can be applied to different local regions in HDR images. The use of the total variation in (15) for the nonuniform weight map can avoid the occurrence of "halo" artifacts which often occur in local mapping methods.

**Fig. 3** Synagogue scene

(a) The proposed method with $\mu = 1$

(b) A uniform weight map $\mathbf{W} = \mathbf{1}$

(c) Drago's method [16]

(d) Ashikhmin's method [11]

(e) Banterle's method [28]

(f) Durand's method [12]

(g) Fattal's method [29]

(h) Reinhard's method [30]

(i) Paris's method [31]

(j) Yee's method [32]

**Fig. 4** The zoomed region of Synagogue scene

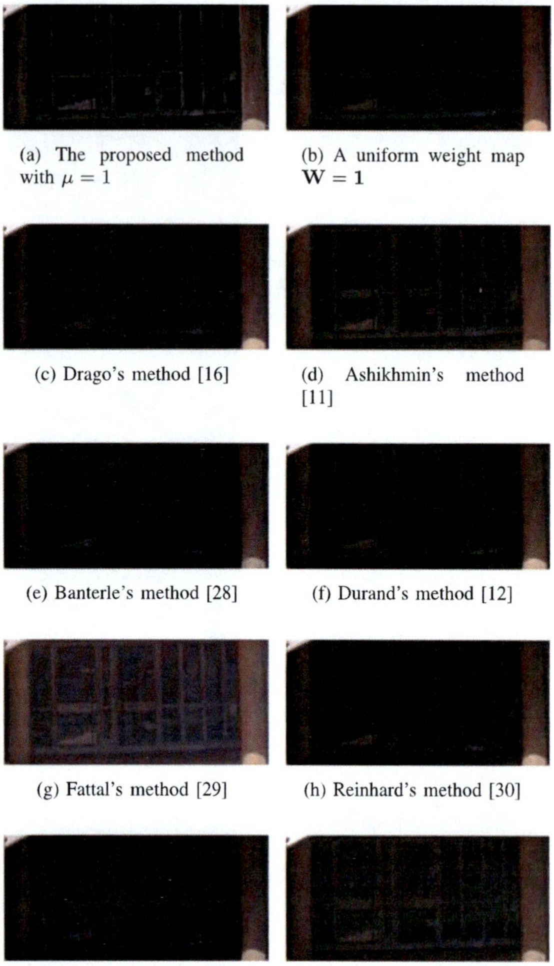

(a) The proposed method with $\mu = 1$

(b) A uniform weight map $\mathbf{W} = \mathbf{1}$

(c) Drago's method [16]

(d) Ashikhmin's method [11]

(e) Banterle's method [28]

(f) Durand's method [12]

(g) Fattal's method [29]

(h) Reinhard's method [30]

(i) Paris's method [31]

(j) Yee's method [32]

## 4 Concluding Remarks

In this paper, we proposed a variational method for handling tone mapping problems. We combine the step of compressing the luminance and the step of color rendering into a variational framework by solving a nonuniform weight map for Gamma correction. The energy function to be minimized consists of a fidelity term and an total variation regularization term. The alternating method of multipliers algorithm is applied to solve the minimization problem. The results on several test HDR images are shown to compare with other well-known tone mapping methods. The results have shown that the visibility of fine details in both bright and dark regions by the proposed methods are more clear than other methods. Also the artifacts around the

**Fig. 5** Indoor scene

(a) The proposed method with $\mu = 1$

(b) A uniform weight map $\mathbf{W} = \mathbf{1}$

(c) Drago's method [16]

(d) Ashikhmin's method [11]

(e) Banterle's method [28]

(f) Durand's method [12]

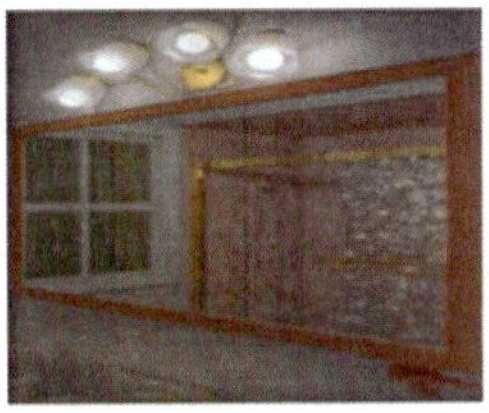

(g) Fattal's method [29]

(h) Reinhard's method [30]

(i) Paris's method [31]

(j) Yee's method [32]

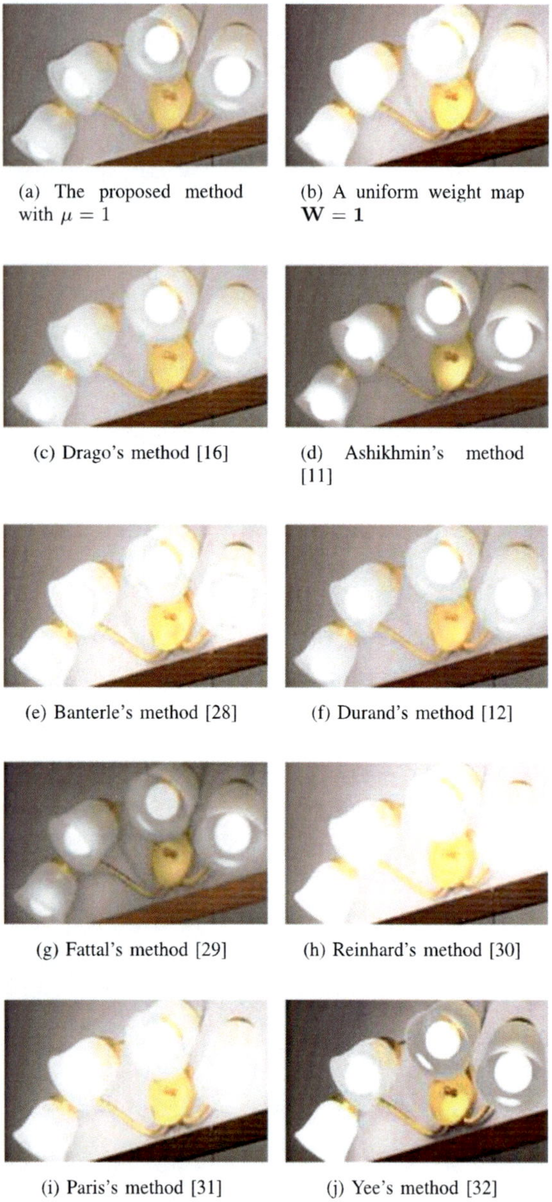

**Fig. 6** The zoomed region of Indoor scene

(a) The proposed method with $\mu = 1$

(b) A uniform weight map $\mathbf{W} = \mathbf{1}$

(c) Drago's method [16]

(d) Ashikhmin's method [11]

(e) Banterle's method [28]

(f) Durand's method [12]

(g) Fattal's method [29]

(h) Reinhard's method [30]

(i) Paris's method [31]

(j) Yee's method [32]

**Fig. 7** Nave scene

(a) The proposed method with $\mu = 1$

(b) A uniform weight map $\mathbf{W} = \mathbf{1}$

(c) Drago's method [16]

(d) Ashikhmin's method [11]

(e) Banterle's method [28]

(f) Durand's method [12]

(g) Fattal's method [29]

(h) Reinhard's method [30]

(i) Paris's method [31]

(j) Yee's method [32]

**Fig. 8** The zoomed region of Nave scene

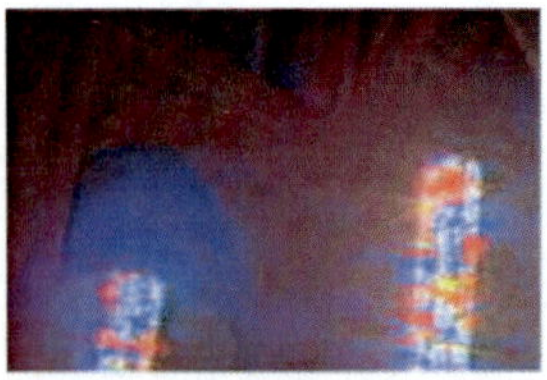

(a) The proposed method with $\mu = 1$

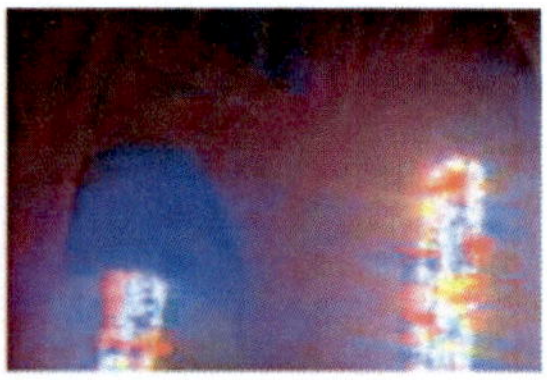

(b) A uniform weight map $\mathbf{W} = \mathbf{1}$

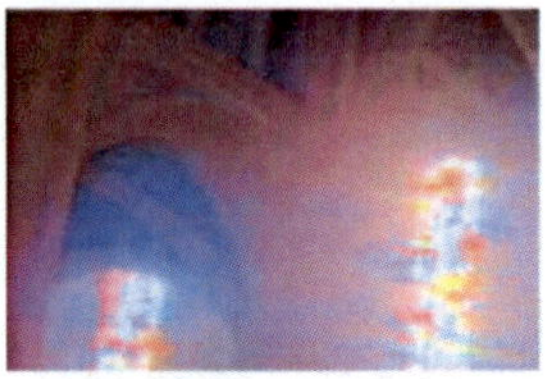

(c) Drago's method [16]

(d) Ashikhmin's method [11]

(e) Banterle's method [28]

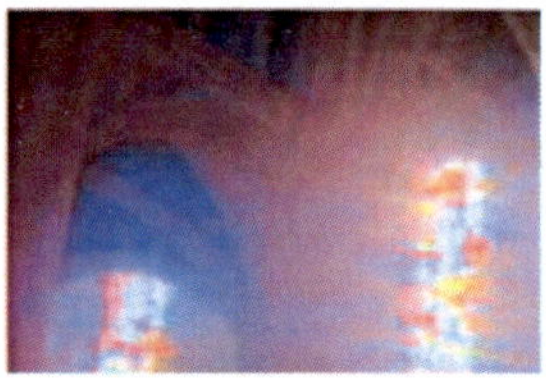

(f) Durand's method [12]

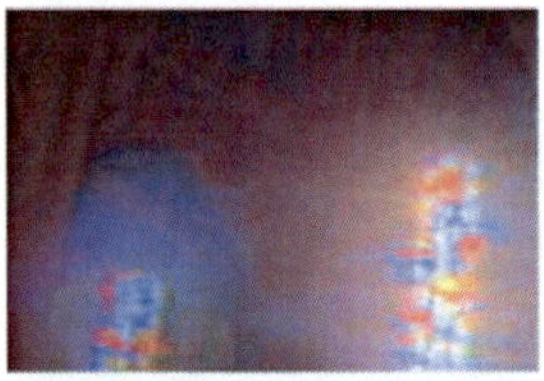

(g) Fattal's method [29]

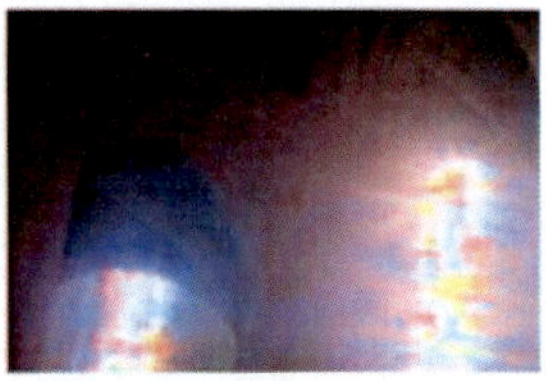

(h) Reinhard's method [30]

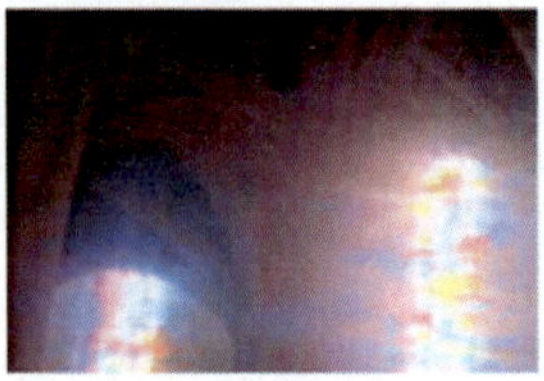

(i) Paris's method [31]

(j) Yee's method [32]

**Fig. 9** Car Park scene

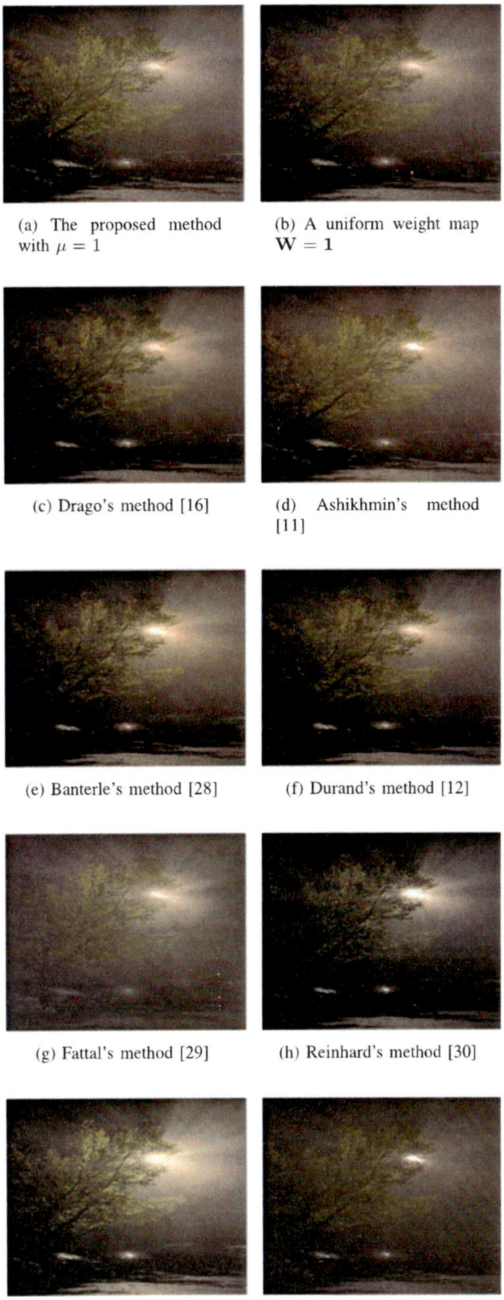

(a) The proposed method with $\mu = 1$

(b) A uniform weight map $\mathbf{W} = \mathbf{1}$

(c) Drago's method [16]

(d) Ashikhmin's method [11]

(e) Banterle's method [28]

(f) Durand's method [12]

(g) Fattal's method [29]

(h) Reinhard's method [30]

(i) Paris's method [31]

(j) Yee's method [32]

**Fig. 10** The zoomed region of Car Park scene

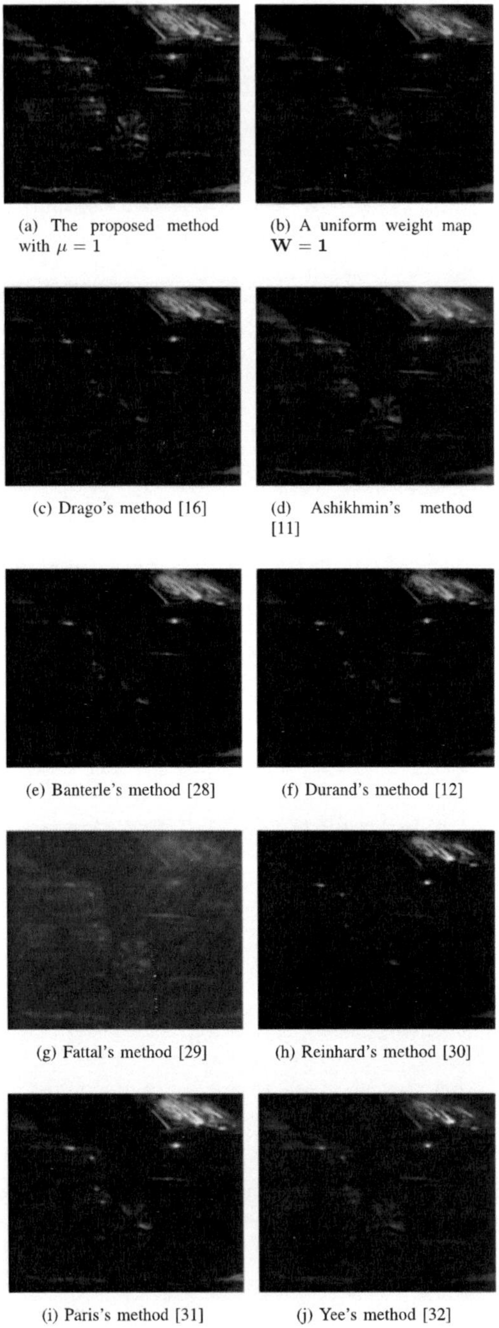

(a) The proposed method with $\mu = 1$

(b) A uniform weight map $\mathbf{W} = \mathbf{1}$

(c) Drago's method [16]

(d) Ashikhmin's method [11]

(e) Banterle's method [28]

(f) Durand's method [12]

(g) Fattal's method [29]

(h) Reinhard's method [30]

(i) Paris's method [31]

(j) Yee's method [32]

**Fig. 11** Belgium scene

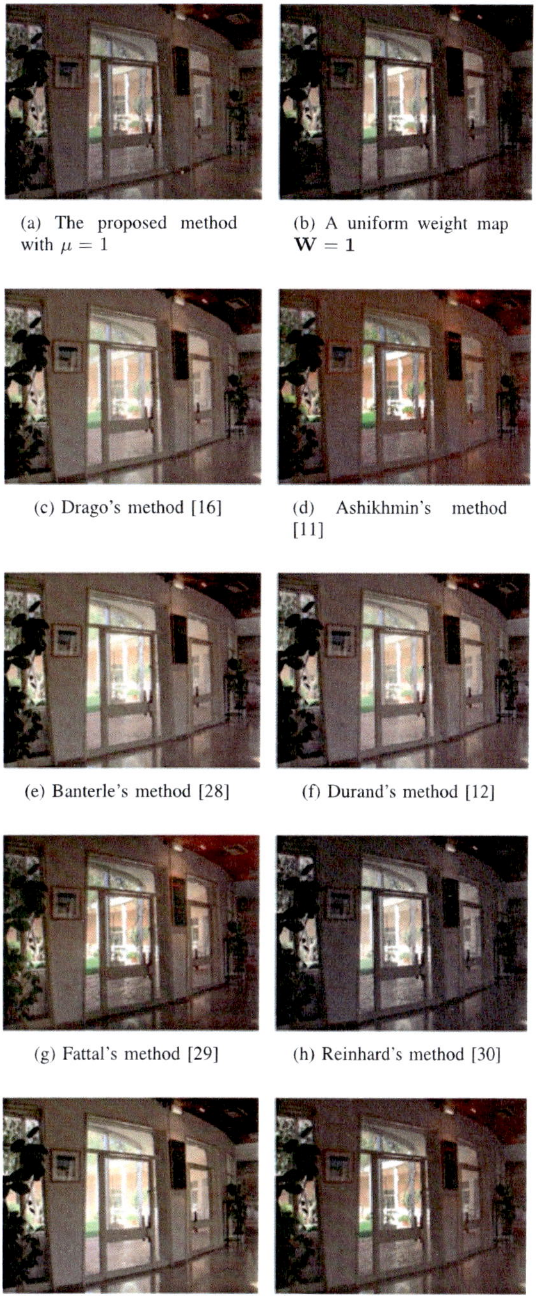

(a) The proposed method with $\mu = 1$

(b) A uniform weight map $\mathbf{W} = \mathbf{1}$

(c) Drago's method [16]

(d) Ashikhmin's method [11]

(e) Banterle's method [28]

(f) Durand's method [12]

(g) Fattal's method [29]

(h) Reinhard's method [30]

(i) Paris's method [31]

(j) Yee's method [32]

**Fig. 12** The zoomed region of Belgium scene

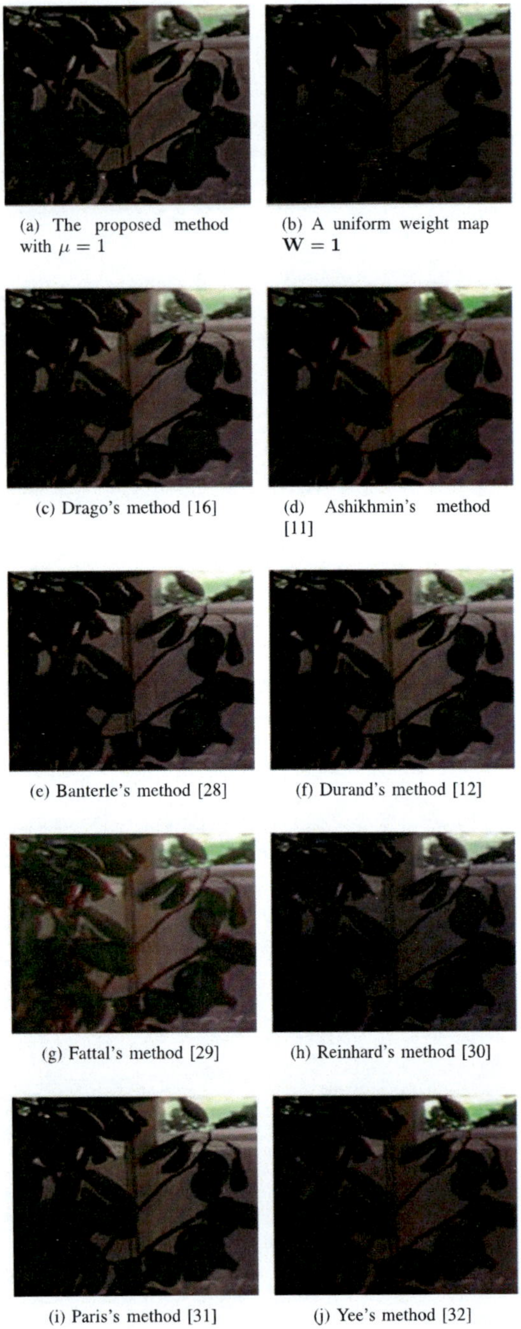

(a) The proposed method with $\mu = 1$

(b) A uniform weight map $\mathbf{W} = \mathbf{1}$

(c) Drago's method [16]

(d) Ashikhmin's method [11]

(e) Banterle's method [28]

(f) Durand's method [12]

(g) Fattal's method [29]

(h) Reinhard's method [30]

(i) Paris's method [31]

(j) Yee's method [32]

edges usually occur in local tone mapping methods can be avoided by using the total variation regularization.

As a future research work, long range dependency among pixels is also an important prior, non-local total variation may be considered in the regularization item. Moreover, deep learning techniques have shown prominent performance in many image processing problems including tone mapping. Some CNN based methods have been proposed, see [31–34]. It would be interesting to study an approach to integrate the proposed method with CNN to further improve the performance [35, 36].

## References


1. R. Erik, W. Heidrich, P. Debevec, S. Pattanaik, G. Ward, K. Myszkowski, *High Dynamic Range Imaging: Acquisition, Display, and Image-Based Lighting* (Morgan Kaufman, 2010)
2. M. Anderson, R. Motta, S. Chandrasekar, M. Stokes, Proposal for a standard default color space for the internet-sRGB. IEEE Conf. Image Process. **1**, 238–245 (1996)
3. G. Wyszecki, W.S. Stiles, Color Science, vol. 8, Wiley, New York (1982)
4. J. Tumblin, H. Rushmeierr, Tone reproduction for realistic images. Comput. Graph. Appl. **6**, 42–48 (1993)
5. J. Stevens, S. Stevens, Brightness function: effects of adaptation. JOSA **53** (3), 375–385 (1963)
6. G. Ward, A contrast-based scalefactor for luminance display. Graph. gems IV, 415–421 (1994)
7. C. Schlick, Quantization techniques for visualization of high dynamic range pictures. Photorealistic Rendering Techniques, 7–20 (1995)
8. K. Chiu, M. Herf, P. Shirley, S. Swamy, C. Wang, K. Zimmerman, Spatially nonuniform scaling functions for high contrast images, in *Graphics Interface*, 245–245. Canadian Information Processing Society (1993)
9. M. Ashikhmin, A tone mapping algorithm for high contrast images, in *Proceedings of the 13th Eurographics Workshop on Rendering*, 145–156, Eurographics Association (2002)
10. J. Tumblin, J. Hodgins, B. Guenter, Two methods for display of high contrast images. ACM Trans. Graph. (TOG) **18**(1), 56–94 (1999)
11. F. Durand, J. Dorsey, Fast bilateral filtering for the display of high-dynamic-range images. ACM Trans. Graph. (TOG) **21**(3), 257–266 (2002) (ACM)
12. C. Tomasi, R. Manduchi, Bilateral filtering for gray and color images, *Sixth IEEE International Conference on Computer Vision* (1988)
13. P. Choudhury, J. Tumblin, The trilateral filter for high contrast images and meshes, *ACM SIGGRAPH* (2005)
14. F. Dragoa, W. Martensb, K. Myszkowskic, N. Chiba, *Design of a Tone Mapping Operator for High-Dynamic Range Images Based upon Psychophysical Evaluation and Preference Mapping* (International Society for Optics and Photonics, Electronic Imaging, 2003)
15. E. Lan, J. McCann, Lightness and retinex theory. JOSA **61**, 1–11 (1971)
16. L. Meylan, S. Susstrunk, High dynamic range image rendering with a retinex-based adaptive filter. IEEE Trans. Image Process. **15**, 2820–2830 (2006)
17. K. Kyungman, J. Bae, J. Kim, Natural HDR image tone mapping based on retinex. IEEE Trans. Consumer Electron. **57**, 1807–1814 (2011)
18. G. Larson, R. Holly, C. Piatko, A visibility matching tone reproduction operator for high dynamic range scenes. IEEE Trans. Visual. Comput. Graph. **3**(4), 291–306 (1997)
19. J. Duan, M. Bressan, C. Dance, G. Qiu, Tone-mapping high dynamic range images by novel histogram adjustment. Pattern Recogn. **43**(5), 1847–1862 (2010)
20. G. Qiu, J. Duan, An optimal tone reproduction curve operator for the display of high dynamic range images. *IEEE International Symposium on Circuits and Systems* (2005)

21. G. Qiu, J. Guan, M. Chen, Tone mapping for HDR image using optimization a new closed form solution, *18th International Conference on Pattern Recognition*, vol. 1, pp. 996–999 (2006)
22. Q. Shan, J. Jia, M. Brown, Globally optimized linear windowed tone mapping. IEEE Trans. Visual. Comput. Graph. **16**, 663–675 (2009)
23. P. Gouras, *Principles of Neural Science*, 3rd ed., E. R. Kandel, J. H. Schwartz, and T. M. Jessell, Eds., Prentice-Hall, New York (1991)
24. M. Ng, F. Wang, X. Yuan, Inexact alternating direction methods for image recovery. SIAM J. Scientific Comput. **33**, 1643–1668 (2011)
25. F. Drago, K. Myszkowski, T. Annen, N. Chiba, Adaptive logarithmic mapping for displaying high contrast scenes. Comput. Graph. Forum **22**(3), 419–426. Blackwell Publishing, Inc. (2003)
26. F. Banterle, A. Artusi, K. Debattista, A. Chalmers, *Advanced High Dynamic Range Imaging: Theory and Practice*. CRC Press (2011)
27. R. Fattal, D. Lischinski, M. Werman, Gradient domain high dynamic range compression. ACM Trans. Graph. (TOG) **21**(3), 249–256 (2002)
28. E. Reinhard, M. Stark, P. Shirley, J. Ferwerda, Photographic tone reproduction for digital images. ACM Trans. Graph. (TOG) **21**(3), 267–276 (2002)
29. S. Paris, F. Durand, A fast approximation of the bilateral filter using a signal processing approach. Int. J. Comput. Vis. **81**(1), 24–52 (2009)
30. Y. Yee, S. Pattanaik, Segmentation and adaptive assimilation for detail-preserving display of high-dynamic range images. Visual Comput. **19**(7–8), 457–466 (2003)
31. Y. Endo, Y. Kanamori, J. Mitani, Deep reverse tone mapping. ACM Trans. Graph. **36**, 177–1 (2017)
32. X. Hou, J. Duan, G. Qiu, Deep feature consistent deep image transformations: Downscaling, decolorization and hdr tone mapping, arXiv preprint arXiv:1707.09482 (2017)
33. J. Lee, K. Sunkavalli, Z. Lin, X. Shen, I. Kweon, Automatic content-aware color and tone stylization, *Proceedings of the IEEE Conference on Computer Vision and Pattern Recognition*, pp. 2470–2478 (2016)
34. S. Ning, H. Xu, L. Song, R. Xie, W. Zhang, Learning an inverse tone mapping network with a generative adversarial regularizer, in *2018 IEEE International Conference on Acoustics, Speech and Signal Processing (ICASSP)*, pp. 1383–1387 (2018)
35. K. Plataniotis, A. Venetsanopoulos, *Color Image Processing and Applications*, Springer (2000)
36. J. Tumblin, G. Turk, LCIS: a boundary hierarchy for detail-preserving contrast reduction, in *Proceedings of the 26th Annual Conference on Computer Graphics and Interactive Techniques*, 83–90. ACM Press/Addison-Wesley Publishing Co. (1999)

# On the Optimal Proximal Parameter of an ADMM-like Splitting Method for Separable Convex Programming

**Bingsheng He and Xiaoming Yuan**

**Abstract** We proposed an ADMM-like splitting method in [11] for solving convex minimization problems with linear constraints and multi-block separable objective functions. Its proximal parameter is required to be sufficiently large to theoretically ensure the convergence, despite that a smaller value of this parameter is preferred for numerical acceleration. Empirically, this method has been applied to solve various applications with relaxed restrictions on the parameter, yet no rigorous theory is available for guaranteeing the convergence. In this paper, we identify the optimal (smallest) proximal parameter for this method and clarify some ambiguity in selecting this parameter for implementation. For succinctness, we focus on the case where the objective function is the sum of three functions and show that the optimal proximal parameter is 0.5. This optimal proximal parameter generates positive indefiniteness in the regularization of the subproblems, and thus its convergence analysis is significantly different from those for existing methods of the same kind in the literature, which all require positive definiteness (or positive semi-definiteness plus additional assumptions) of the regularization. We establish the convergence and estimate the convergence rate in terms of iteration complexity for the improved method with the optimal proximal parameter.

**Keywords** Convex programming · Splitting method · Positive indefinite proximal regularization · Convergence analysis

Bingsheng He—He was supported by the NSFC Grant 11471156. // Xiaoming Yuan—He was supported by the General Research Fund from Hong Kong Research Grants Council: 12300317.

B. He
Department of Mathematics, Southern University of Science and Technology, Shenzhen, China
e-mail: hebma@nju.edu.cn

Department of Mathematics, Nanjing University, Nanjing, China

X. Yuan (✉)
Department of Mathematics, The University of Hong Kong, Pok Fu Lam, Hong Kong
e-mail: xmyuan@hku.hk



X.-C. Tai et al. (eds.), *Mathematical Methods in Image Processing and Inverse Problems*, Springer Proceedings in Mathematics & Statistics 360,
https://doi.org/10.1007/978-981-16-2701-9_8

## 1 Introduction

Our purpose is finding the optimal (smallest) proximal parameter for the splitting method in [11] for separable convex programming models. To expose our main idea and technique more clearly, we focus on the special convex minimization problem with linear constraints and a separable objective function that can be represented as the sum of three functions without coupled variables:

$$\min\{\theta_1(x) + \theta_2(y) + \theta_3(z) \mid Ax + By + Cz = b,\ x \in \mathcal{X}, y \in \mathcal{Y}, z \in \mathcal{Z}\}, \quad (1)$$

where $A \in \Re^{m\times n_1}$, $B \in \Re^{m\times n_2}$, $C \in \Re^{m\times n_3}$; $b \in \Re^m$; $\mathcal{X} \subset \Re^{n_1}$, $\mathcal{Y} \subset \Re^{n_2}$ and $\mathcal{Z} \subset \Re^{n_3}$ are closed convex sets; and $\theta_i : \Re^{n_i} \to \Re$ $(i = 1, 2, 3)$ are closed convex but not necessarily smooth functions. Such a model may arise from a concrete application in which one of the functions represents a data-fidelity term while the other two account for various regularization terms. We refer to, e.g., [16, 20–23], for some applications of (1). The solution set of (1) is assumed to be nonempty throughout.

To recall the splitting method in [11] for the model (1), we start from the augmented Lagrangian method (ALM) that was originally proposed in [15, 18]. Let the Lagrangian and augmented Lagrangian functions of (1) be given, respectively, by

$$L(x, y, z, \lambda) = \theta_1(x) + \theta_2(y) + \theta_3(y) - \lambda^T(Ax + By + Cz - b), \quad (2)$$

and

$$\mathcal{L}_\beta(x, y, z, \lambda) = \theta_1(x) + \theta_2(y) + \theta_3(z) - \lambda^T(Ax + By + Cz - b) + \frac{\beta}{2}\|Ax + By + Cz - b\|^2. \quad (3)$$

In (2) and (3), $\lambda \in \Re^m$ is the Lagrange multiplier; and in (3), $\beta > 0$ is the penalty parameter. When the three-block separable convex minimization model (1) is purposively regarded as a generic convex minimization model and its objective function is treated as a whole, the ALM in [15, 18] can be applied directly and the resulting iterative scheme is

$$\begin{cases} (x^{k+1}, y^{k+1}, z^{k+1}) = \arg\min\{\mathcal{L}_\beta(x, y, z, \lambda^k) \mid x \in \mathcal{X}, y \in \mathcal{Y}, z \in \mathcal{Z}\}, & \text{(4a)} \\ \lambda^{k+1} = \lambda^k - \beta(Ax^{k+1} + By^{k+1} + Cz^{k+1} - b). & \text{(4b)} \end{cases}$$

If two functions in the objective are treated together and two variables in the constraints are grouped accordingly, the alternating direction method of multipliers (ADMM) in [5] can also be directly applied to (1). The resulting iterative scheme reads as

$$\begin{cases} x^{k+1} = \arg\min\{\mathcal{L}_\beta(x, y^k, z^k, \lambda^k) \mid x \in \mathcal{X}\}, & \text{(5a)} \\ (y^{k+1}, z^{k+1}) = \arg\min\{\mathcal{L}_\beta(x^{k+1}, y, z, \lambda^k) \mid y \in \mathcal{Y}, z \in \mathcal{Z}\}, & \text{(5b)} \\ \lambda^{k+1} = \lambda^k - \beta(Ax^{k+1} + By^{k+1} + Cz^{k+1} - b). & \text{(5c)} \end{cases}$$

Unless the functions and/or coefficient matrices in (1) are special enough, direct applications of the ALM (1.4) and the ADMM (1.5) usually are not preferred because the $(x, y, z)$-subproblem in (1.5b) and $(y, z)$-subproblem in (1.5b) may still be too difficult (even when the functions $\theta_i$ per se are relatively easy). Therefore, generally the three-block model (1) should not be treated as a one-block or two-block case and the ALM (1.4) or ADMM (1.5) should not be applied directly.

On the other hand, for specific applications of the model (1), functions in its objective usually have their own physical explanations and mathematical properties. Thus, it is usually necessary to treat them individually to design more efficient algorithms. More accurately, we are interested in such an algorithm that handles these functions $\theta_i$ individually in its iterative scheme. A natural idea is to split the subproblem in the original ALM (1.4) in the Jacobian or Gaussian manner; the corresponding schemes are as follows:

$$\begin{cases} x^{k+1} = \arg\min\{\mathcal{L}_\beta(x, y^k, z^k, \lambda^k) \mid x \in \mathcal{X}\}, \\ y^{k+1} = \arg\min\{\mathcal{L}_\beta(x^k, y, z^k, \lambda^k) \mid y \in \mathcal{Y}\}, \\ z^{k+1} = \arg\min\{\mathcal{L}_\beta(x^k, y^k, z, \lambda^k) \mid z \in \mathcal{Z}\}, \\ \lambda^{k+1} = \lambda^k - \beta(Ax^{k+1} + By^{k+1} + Cz^{k+1} - b), \end{cases} \tag{6}$$

and

$$\begin{cases} x^{k+1} = \arg\min\{\mathcal{L}_\beta(x, y^k, z^k, \lambda^k) \mid x \in \mathcal{X}\}, \\ y^{k+1} = \arg\min\{\mathcal{L}_\beta(x^{k+1}, y, z^k, \lambda^k) \mid y \in \mathcal{Y}\}, \\ z^{k+1} = \arg\min\{\mathcal{L}_\beta(x^{k+1}, y^{k+1}, z, \lambda^k) \mid z \in \mathcal{Z}\}, \\ \lambda^{k+1} = \lambda^k - \beta(Ax^{k+1} + By^{k+1} + Cz^{k+1} - b). \end{cases} \tag{7}$$

All the subproblems in (6) and (7) are easier than the original problem (1); only one function in its objective and a quadratic term are involved in the $x$-, $y$-, $z$-subproblems. But, as shown in [1, 8], neither of the schemes (6) and (7) is necessarily convergent. Therefore, although schemes such as (6) and (7) can be easily generated, the lack of convergence may require more meticulous theoretical study and algorithmic design techniques for the three-block case (1). The results in [1, 8] also justify that designing augmented-Lagrangian-based splitting algorithms for the three-block case (1) is significantly different from that for the one- or two-block case; and they need to be discussed separately despite that there is a rich literature of the ALM and ADMM.

Despite of their lack of convergence, the schemes (6) and (7) may empirically work well, see, e.g., [20, 22, 23]. It is thus interesting to design an augmented-Lagrangian-based splitting method whose iterative scheme is analogous to (6), (7), or a fused one of both, while its theoretical convergence and empirical efficiency can be both ensured. The method in [11] is such one; its iterative scheme for (1) reads as

$$\begin{cases} x^{k+1} = \arg\min\{\mathcal{L}_\beta(x, y^k, z^k, \lambda^k) \mid x \in \mathcal{X}\}, & \text{(8a)} \\ \lambda^{k+\frac{1}{2}} = \lambda^k - \beta(Ax^{k+1} + By^k + Cz^k - b), & \text{(8b)} \\ \begin{cases} y^{k+1} = \text{argmin}\{\theta_2(y) - (\lambda^{k+\frac{1}{2}})^T By + \frac{\mu\beta}{2}\|B(y - y^k)\|^2 \mid y \in \mathcal{Y}\}, \\ z^{k+1} = \text{argmin}\{\theta_3(z) - (\lambda^{k+\frac{1}{2}})^T Cz + \frac{\mu\beta}{2}\|C(z - z^k)\|^2 \mid z \in \mathcal{Z}\}, \end{cases} & \text{(8c)} \\ \lambda^{k+1} = \lambda^k - \beta(Ax^{k+1} + By^{k+1} + Cz^{k+1} - b), & \text{(8d)} \end{cases}$$

where the parameter $\mu$ is required to be $\mu \geq 2$ in [11]. The scheme (1.8) has the simplicity in sense of that each of the $x$-, $y$-, and $z$-subproblems involves just one function from (1) in its objective. Its efficiency has been verified in [11] by some sparse and low-rank models and image inpainting problems. Also, it was used in [2] for solving a dimensionality reduction problem on physical space.

It is easy to see that the scheme (1.8) can be rewritten as

$$\begin{cases} x^{k+1} = \arg\min\{\mathcal{L}_\beta(x, y^k, z^k, \lambda^k) \mid x \in \mathcal{X}\}, & \text{(9a)} \\ \begin{cases} y^{k+1} = \arg\min\{\mathcal{L}_\beta(x^{k+1}, y, z^k, \lambda^k) + \frac{\tau\beta}{2}\|B(y - y^k)\|^2 \mid y \in \mathcal{Y}\}, \\ z^{k+1} = \arg\min\{\mathcal{L}_\beta(x^{k+1}, y^k, z, \lambda^k) + \frac{\tau\beta}{2}\|C(z - z^k)\|^2 \mid z \in \mathcal{Z}\}, \end{cases} & \text{(9b)} \\ \lambda^{k+1} = \lambda^k - \beta(Ax^{k+1} + By^{k+1} + Cz^{k+1} - b), & \text{(9c)} \end{cases}$$

with $\tau = \mu - 1$ and thus $\tau \geq 1$ as shown in [11]. The scheme (1.9) shows more clearly that it is a mixture of the augmented-Lagrangian-based splitting schemes (6) and (7), in which the $x$- and $(y, z)$-subproblems are updated in the alternating order while the $(y, z)$-subproblem is further splitted in parallel so that parallel computation can be implemented to the resulting $y$- and $z$-subproblems. Recall the lack of convergence of (6) and (7). Thus, it is necessary to regularize the splitted $y$- and $z$-subproblems appropriately in (1.9) to ensure the convergence. Indeed, the terms $\frac{\tau\beta}{2}\|B(y - y^k)\|^2$ and $\frac{\tau\beta}{2}\|C(z - z^k)\|^2$ in (1.9) can be regarded as proximal regularization terms with $\tau$ as the proximal parameter.

On the other hand, with fixed $\beta$, the proximal parameter $\tau$ determines the weight of the proximal terms in the subproblems (1.9b) and its reciprocal plays the role of step size for an algorithm implemented internally to solve the subproblems (1.9b). We hence prefer smaller values of $\tau$ whenever the convergence of (1.9) can be theoretically guaranteed. As mentioned, in [11], we have shown that the condition $\tau \geq 1$ is sufficient to ensure the convergence of (1.9). While, numerically, as shown in [11] and also in [2] (see Section V, Part B, Pages 3247–3248), it has been observed that values very close to 1 are preferred for $\tau$. For example, $\mu = 2.01$, i.e., $\tau = 1.01$, was recommended in [11] and used in [2] to result in faster convergence. This raises the necessity of seeking the optimal (smallest) value of $\tau$ that can ensure the convergence of (1.9). The main purpose of this paper is to rigorously prove that the optimal value of $\tau$ is 0.5 for the method (1.9). That is, any $\tau > 0.5$ ensures the convergence of (1.9) yet any $\tau \in (0, 0.5)$ yields divergence.

Note that, because of our analysis in [1], without loss of the generality, we can just assume $\beta \equiv 1$. That is, the augmented Lagrangian function defined in (3) is reduced to

$$\begin{aligned}\mathcal{L}(x, y, z, \lambda) = \theta_1(x) + \theta_2(y) + \theta_3(z) \\ - \lambda^T(Ax + By + Cz - b) + \frac{1}{2}\|Ax + By + Cz - b\|^2; \end{aligned} \quad (10)$$

and the iterative scheme of (1.9) is now simplified as

$$\begin{cases} x^{k+1} = \arg\min\big\{\mathcal{L}(x, y^k, z^k, \lambda^k) \mid x \in \mathcal{X}\big\}, & (11a) \\ \begin{cases} y^{k+1} = \arg\min\big\{\mathcal{L}(x^{k+1}, y, z^k, \lambda^k) + \frac{\tau}{2}\|B(y - y^k)\|^2 \mid y \in \mathcal{Y}\big\}, \\ z^{k+1} = \arg\min\big\{\mathcal{L}(x^{k+1}, y^k, z, \lambda^k) + \frac{\tau}{2}\|C(z - z^k)\|^2 \mid z \in \mathcal{Z}\big\}, \end{cases} & (11b) \\ \lambda^{k+1} = \lambda^k - (Ax^{k+1} + By^{k+1} + Cz^{k+1} - b). & (11c) \end{cases}$$

The rest of this paper is organized as follows. We recall some preliminaries in Sect. 2. In Sect. 3, we show why positive indefiniteness occurs in the proximal regularization for the scheme (1.11) when $\tau > 0.5$. Then, we provide an explanation in the prediction-correction framework for (1.11) in Sect. 4; and focus on analyzing an important quadratic term in Sect. 5 that is the key for conducting convergence analysis for (1.11). The convergence of (1.11) with $\tau > 0.5$ is proved in Sect. 6; and the divergence of (1.11) with $\tau \in (0, 0.5)$ is shown in Sect. 7 by an example. We estimate the worst-case convergence rate in terms of iteration complexity for the scheme (1.11) in Sect. 8. Finally, we make some conclusions in Sect. 9.

## 2 Preliminaries

In this section, we recall some preliminary results for further analysis. First of all, a pair of $\big((x^*, y^*, z^*), \lambda^*\big)$ is called a saddle point of the Lagrangian function defined in (2) if it satisfies the inequalities

$$L_{\lambda\in\Re^m}(x^*, y^*, z^*, \lambda) \le L(x^*, y^*, z^*, \lambda^*) \le L_{x\in\mathcal{X}, y\in\mathcal{Y}, z\in\mathcal{Z}}(x, y, z, \lambda^*).$$

Or, we can rewrite these inequalities as

$$\begin{cases} x^* = \arg\min\{L(x, y^*, z^*, \lambda^*) \mid x \in \mathcal{X}\}, \\ y^* = \arg\min\{L(x^*, y, z^*, \lambda^*) \mid y \in \mathcal{Y}\}, \\ z^* = \arg\min\{L(x^*, y^*, z, \lambda^*) \mid z \in \mathcal{Z}\}, \\ \lambda^* = \arg\max\{L(x^*, y^*, z^*, \lambda) \mid \lambda \in \Re^m\}. \end{cases} \quad (12)$$

Indeed, a saddle point of the Lagrangian function defined in (2) can also be characterized by the following variational inequality:

$$\begin{cases} x^* \in \mathcal{X}, & \theta_1(x) - \theta_1(x^*) + (x - x^*)^T(-A^T\lambda^*) \geq 0, \ \forall\, x \in \mathcal{X}, \\ y^* \in \mathcal{Y}, & \theta_2(y) - \theta_2(y^*) + (y - y^*)^T(-B^T\lambda^*) \geq 0, \ \forall\, y \in \mathcal{Y}, \\ z^* \in \mathcal{Z}, & \theta_3(z) - \theta_3(z^*) + (z - z^*)^T(-C^T\lambda^*) \geq 0, \ \forall\, z \in \mathcal{Z}, \\ \lambda^* \in \Re^m, & (\lambda - \lambda^*)^T(Ax^* + By^* + Cz^* - b) \geq 0, \ \forall\, \lambda \in \Re^m. \end{cases} \tag{13}$$

We call $(x, y, z)$ and $\lambda$ the primal and dual variables, respectively.

The optimality condition of the model (1) can be characterized by the monotone variational inequality:

$$w^* \in \Omega, \quad \theta(u) - \theta(u^*) + (w - w^*)^T F(w^*) \geq 0, \quad \forall\, w \in \Omega, \tag{14a}$$

where

$$u = \begin{pmatrix} x \\ y \\ z \end{pmatrix}, \ \theta(u) = \theta_1(x) + \theta_2(y) + \theta_3(z), \ w = \begin{pmatrix} x \\ y \\ z \\ \lambda \end{pmatrix}, \ F(w) = \begin{pmatrix} -A^T\lambda \\ -B^T\lambda \\ -C^T\lambda \\ Ax + By + Cz - b \end{pmatrix} \tag{14b}$$

and

$$\Omega = \mathcal{X} \times \mathcal{Y} \times \mathcal{Z} \times \Re^m.$$

We denote by $\Omega^*$ the solution set of (14). Note that the operator $F$ in (14b) is affine with a skew-symmetric matrix and thus we have

$$(w - \bar{w})^T(F(w) - F(\bar{w})) = 0, \quad \forall\, w, \bar{w}. \tag{15}$$

## 3 The Positive Indefiniteness of (1.11) with $\tau > 0.5$

In this section, we revisit the scheme (1.11) from the variational inequality perspective; and show that it can be represented as a proximal version of the direct application of ADMM (1.5) but the proximal regularization term is not positive definite for the case of $\tau > 0.5$. The positive indefiniteness of the proximal regularization excludes the application of a vast set of known convergence results in the literature of ADMM and its proximal versions, because they all require positive definiteness or semi-definiteness (plus additional assumptions on the model (1)) for the proximal regularization term to validate the convergence analysis.

Let us first take a look at the optimality conditions of the subproblems in (1.11). Note that the subproblem (1.11b) are specified as

$$y^{k+1} = \arg\min\{\theta_2(y) - y^T B\lambda^k + \frac{1}{2}\|Ax^{k+1} + By + Cz^k - b\|^2 + \frac{\tau}{2}\|B(y - y^k)\|^2 \mid y \in \mathcal{Y}\}, \tag{16a}$$

and

$$z^{k+1} = \arg\min\{\theta_3(z) - z^T C\lambda^k + \frac{1}{2}\|Ax^{k+1} + By^k + Cz - b\|^2 + \frac{\tau}{2}\|C(z - z^k)\|^2 \mid z \in \mathcal{Z}\}. \tag{16b}$$

Thus, the optimality condition of the $y$-subproblem in (1.11b) can be written as $y^{k+1} \in \mathcal{Y}$ and

$$\theta_2(y) - \theta_2(y^{k+1}) + (y - y^{k+1})^T \begin{pmatrix} -B^T\lambda^k + B^T(Ax^{k+1} + By^{k+1} + Cz^k - b) \\ +\tau B^T B(y^{k+1} - y^k) \end{pmatrix} \geq 0, \ \forall y \in \mathcal{Y};$$

or equivalently: $y^{k+1} \in \mathcal{Y}$ and

$$\theta_2(y) - \theta_2(y^{k+1}) + (y - y^{k+1})^T \begin{pmatrix} -B^T\lambda^k + B^T(Ax^{k+1} + By^{k+1} + Cz^{k+1} - b) \\ \tau B^T B(y^{k+1} - y^k) - B^T C(z^{k+1} - z^k) \end{pmatrix} \geq 0, \ \forall y \in \mathcal{Y}. \tag{17a}$$

Similarly, the optimality condition of the $z$-subproblem in (1.11b) can be written as $z^{k+1} \in \mathcal{Z}$ and

$$\theta_3(z) - \theta_3(z^{k+1}) + (z - z^{k+1})^T \begin{pmatrix} -C^T\lambda^k + C^T(Ax^{k+1} + By^{k+1} + Cz^{k+1} - b) \\ -C^T B(y^{k+1} - y^k) + \tau C^T C(z^{k+1} - z^k) \end{pmatrix} \geq 0, \ \forall z \in \mathcal{Z}. \tag{17b}$$

Then, with (1.11c), we can rewrite the inequalities (17a) and (17b) as $(y^{k+1}, z^{k+1}) \in \mathcal{Y} \times \mathcal{Z}$ and

$$\begin{pmatrix} \theta_2(y) - \theta_2(y^{k+1}) \\ \theta_3(z) - \theta_3(z^{k+1}) \end{pmatrix} + \begin{pmatrix} y - y^{k+1} \\ z - z^{k+1} \end{pmatrix}^T \left\{ \begin{pmatrix} -B^T\lambda^{k+1} \\ -C^T\lambda^{k+1} \end{pmatrix} + D_0 \begin{pmatrix} y^{k+1} - y^k \\ z^{k+1} - z^k \end{pmatrix} \right\} \geq 0, \ \forall (y, z) \in \mathcal{Y} \times \mathcal{Z}, \tag{18}$$

where

$$D_0 = \begin{pmatrix} \tau B^T B & -B^T C \\ -C^T B & \tau C^T C \end{pmatrix}. \tag{19}$$

Obviously, $D_0$ is positive semidefinite and indefinite when $\tau \geq 1$ and $\tau \in (0, 1)$, respectively.

Then, it is easy to see that the scheme (1.11) can be rewritten as

$$\begin{cases} x^{k+1} \;=\; \arg\min\{\mathcal{L}(x, y^k, z^k, \lambda^k) \mid x \in \mathcal{X}\}, & (20a)\\ \begin{pmatrix} y^{k+1} \\ z^{k+1} \end{pmatrix} = \arg\min\left\{\mathcal{L}(x^{k+1}, y, z, \lambda^k) + \frac{1}{2}\left\| \begin{matrix} y - y^k \\ z - z^k \end{matrix} \right\|_{D_0}^2 \;\middle|\; (y, z) \in \mathcal{Y} \times \mathcal{Z}\right\}, & (20b)\\ \lambda^{k+1} = \lambda^k - (Ax^{k+1} + By^{k+1} + Cz^{k+1} - b), & (20c) \end{cases}$$

Comparing (3.5b) with (1.5b) (note that $\beta = 1$), we see that the scheme (1.11) can be symbolically represented as a proximal version of (1.5) in which the $(y, z)$-subproblem is proximally regularized by a proximal term. But the difficulty is that $D_0$ defined in (19) is positive indefinite when $\tau \in (0.5, 1)$. Indeed, our analysis in [11] requires $\tau \geq 1$ and thus the positive semidefiniteness of $D_0$ is ensured. For this case, the convergence analysis is relatively easy because it can follow some techniques used for the proximal point algorithm which is originated from [17, 19]. For the case where $\tau$ is relaxed to $\tau > 0.5$ and hence the matrix $D_0$ in (19) is positive indefinite, the analysis in [11] and other literatures is not applicable and more sophisticated techniques are needed for proving the convergence of the scheme (1.11) with $\tau > 0.5$.

## 4 A Prediction-Correction Explanation of (1.11)

In this section, we show that the scheme (1.11) can be expressed by a prediction-correction framework. This prediction-correction explanation is only for the convenience of theoretical analysis and there is no need to follow this prediction-correction framework to implement the scheme (1.11).

In the scheme (1.11), we see that $x^k$ is not needed to generate the next $(k+1)$-th iterate; only $(y^k, z^k, \lambda^k)$ are needed. Thus, we call $x$ the intermediate variable; and $(y, z, \lambda)$ essential variables. To distinguish their roles, accompanied with the notation in (14b), we additionally define the notation

$$v = \begin{pmatrix} y \\ z \\ \lambda \end{pmatrix}, \quad \mathcal{V} = \mathcal{Y} \times \mathcal{Z} \times R^m \quad \text{and} \quad \mathcal{V}^* = \{(y^*, z^*, \lambda^*) \mid (x^*, y^*, z^*, \lambda^*) \in \Omega^*\}. \tag{21}$$

Moreover, we introduce the auxiliary variables $\tilde{w}^k = (\tilde{x}^k, \tilde{y}^k, \tilde{z}^k, \tilde{\lambda}^k)$ defined by

$$\tilde{x}^k = x^{k+1}, \quad \tilde{y}^k = y^{k+1}, \quad \tilde{z}^k = z^{k+1} \quad \text{and} \quad \tilde{\lambda}^k = \lambda^k - (Ax^{k+1} + By^k + Cz^k - b), \tag{22}$$

where $(x^{k+1}, y^{k+1}, z^{k+1})$ is the iterate generated by the scheme (1.11) from the given one $(y^k, z^k, \lambda^k)$. Using these notations, we have

$$\begin{aligned}\lambda^{k+1} &= \lambda^k - (Ax^{k+1} + By^{k+1} + Cz^{k+1} - b)\\ &= [\lambda^k - (Ax^{k+1} + By^k + Cz^k - b)] + B(y^k - y^{k+1}) + C(z^k - z^{k+1})\\ &= \tilde{\lambda}^k + B(y^k - \tilde{y}^k) + C(z^k - \tilde{z}^k).\end{aligned} \tag{23}$$

Now, we interpret the optimality conditions of the subproblems in (1.11) by using the auxiliary variables $\tilde{w}^k$. First, ignoring some constant terms, the subproblem (1.11.a) is equivalent to

$$x^{k+1} = \arg\min\{\theta_1(x) - x^T A\lambda^k + \frac{1}{2}\|Ax + By^k + Cz^k - b\|^2 \mid x \in \mathcal{X}\};$$

and its optimality condition can be rewritten as

$$\tilde{x}^k \in \mathcal{X}, \;\; \theta_1(x) - \theta_1(\tilde{x}^k) + (x - \tilde{x}^k)^T(-A^T\tilde{\lambda}^k) \geq 0, \;\; \forall\, x \in \mathcal{X}. \tag{24a}$$

Using (23), $y^{k+1} = \tilde{y}^k$ and $z^{k+1} = \tilde{z}^k$, the inequalities (17a) and (17b) can be written as

$$\tilde{y}^k \in \mathcal{Y}, \;\; \theta_2(y) - \theta_2(\tilde{y}^k) + (y - \tilde{y}^k)^T\{-B^T\tilde{\lambda}^k + (1+\tau)B^T B(\tilde{y}^k - y^k)\} \geq 0, \;\; \forall\, y \in \mathcal{Y}$$

and

$$\tilde{z}^k \in \mathcal{Z}, \;\; \theta_3(z) - \theta_3(\tilde{z}^k) + (z - \tilde{z}^k)^T\{-C^T\tilde{\lambda}^k + (1+\tau)C^T C(\tilde{z}^k - z^k)\} \geq 0, \;\; \forall\, z \in \mathcal{Z},$$

respectively. Thus, the inequality (18) becomes $(\tilde{y}^k, \tilde{z}^k) \in \mathcal{Y} \times \mathcal{Z}$ and

$$\begin{pmatrix}\theta_2(y) - \theta_2(\tilde{y}^k)\\ \theta_3(z) - \theta_3(\tilde{z}^k)\end{pmatrix} + \begin{pmatrix} y - \tilde{y}^k\\ z - \tilde{z}^k\end{pmatrix}^T\left\{\begin{pmatrix}-B^T\\ -C^T\end{pmatrix}\tilde{\lambda}^k + \right.$$
$$\left. + (1+\tau)\begin{pmatrix}B^TB & 0\\ 0 & C^TC\end{pmatrix}\begin{pmatrix}\tilde{y}^k - y^k\\ \tilde{z}^k - z^k\end{pmatrix}\right\} \geq 0, \;\; \forall (y,z) \in \mathcal{Y} \times \mathcal{Z}. \tag{24b}$$

Note that the equality $\tilde{\lambda}^k = \lambda^k - (Ax^{k+1} + By^k + Cz^k - b)$ in (22) can be written as the variational inequality form

$$\tilde{\lambda}^k \in \Re^m, \;\; (\lambda - \tilde{\lambda}^k)^T\{(A\tilde{x}^k + B\tilde{y}^k + C\tilde{z}^k - b) - B(\tilde{y}^k - y^k) - C(\tilde{z}^k - z^k) + (\tilde{\lambda}^k - \lambda^k)\} \geq 0, \;\; \forall \lambda \in \Re^m. \tag{24c}$$

Therefore, it follows from the inequalities (24a), (24b) and (24c) that the auxiliary variable $\tilde{w}^k = (\tilde{x}^k, \tilde{y}^k, \tilde{z}^k, \tilde{\lambda}^k)$ defined in (22) satisfies the following variational inequality.

**Prediction Step**

$$\tilde{w}^k \in \Omega, \ \theta(u) - \theta(\tilde{u}^k) + (w - \tilde{w}^k)^T F(\tilde{w}^k) \geq (v - \tilde{v}^k)^T Q(v^k - \tilde{v}^k), \quad \forall w \in \Omega, \tag{25a}$$

where

$$Q = \begin{pmatrix} (1+\tau)B^T B & 0 & 0 \\ 0 & (1+\tau)C^T C & 0 \\ -B & -C & I_m \end{pmatrix}. \tag{25b}$$

We call the auxiliary variable $\tilde{w}^k = (\tilde{x}^k, \tilde{y}^k, \tilde{z}^k, \tilde{\lambda}^k)$ as the predictor. Using (23), the update form (1.11c) can be represented as

$$\lambda^{k+1} = \lambda^k - (Ax^{k+1} + By^{k+1} + Cz^{k+1} - b) = \lambda^k - [-B(y^k - \tilde{y}^k) - C(z^k - \tilde{z}^k) + (\lambda^k - \tilde{\lambda}^k)].$$

Recall we define by $v$ in (21) the essential variables for the scheme (1.11). The new essential variables of (1.11), $v^{k+1} = (y^{k+1}, z^{k+1}, \lambda^{k+1})$, are updated by the following scheme:

**Correction Step**

$$v^{k+1} = v^k - M(v^k - \tilde{v}^k), \tag{26a}$$

where

$$M = \begin{pmatrix} I & 0 & 0 \\ 0 & I & 0 \\ -B & -C & I_m \end{pmatrix}. \tag{26b}$$

Overall, the scheme (1.11) can be explained by a prediction-correction framework which generates a predictor characterized by the step (4.5) and then corrects it by the step (4.6). As we shall show, the inequality (4.5) indicates the discrepancy between $\tilde{w}^k$ and a solution point of the variational inequality (14) and it plays an important role in the convergence analysis for the scheme (1.11). Indeed, we can further investigate the inequality (4.5) and derive a new right-hand side that is more preferred for establishing the convergence. For this purpose, let us define a matrix as

$$H = \begin{pmatrix} (1+\tau)B^T B & 0 & 0 \\ 0 & (1+\tau)C^T C & 0 \\ 0 & 0 & I_m \end{pmatrix}, \tag{27}$$

which is positive definite for any $\tau > 0$ when $B$ and $C$ are both full column rank. Then, for the matrices $Q$ and $M$ defined in (4.5b) and (4.6b), respectively, it obviously holds that

$$Q = HM. \tag{28}$$

In the following lemma, we further analyze the right-hand side of (4.5) and show more explicitly the difference of the proof for the convergence of (1.11) with $\tau > 0.5$ from that with $\tau \geq 1$ in [11].

**Theorem 4.1** *Let $\{w^k\}$ be the sequence generated by (1.11) for the problem (1) and $\tilde{w}^k$ be defined by (22). Then, $\tilde{w}^k \in \Omega$ and*

$$\theta(u) - \theta(\tilde{u}^k) + (w - \tilde{w}^k)^T F(w) \geq \frac{1}{2}(\|v - v^{k+1}\|_H^2 - \|v - v^k\|_H^2) + \frac{1}{2}(v^k - \tilde{v}^k)^T G(v^k - \tilde{v}^k), \ \forall w \in \Omega, \tag{29}$$

*where*

$$G = Q^T + Q - M^T H M. \tag{30}$$

***Proof*** Using $Q = HM$ (see (28)) and the relation (4.6a), the right-hand side of (4.5a) can be written as

$$(v - \tilde{v}^k)^T H(v^k - v^{k+1}),$$

and hence we have

$$\theta(u) - \theta(\tilde{u}^k) + (w - \tilde{w}^k)^T F(\tilde{w}^k) \geq (v - \tilde{v}^k)^T H(v^k - v^{k+1}), \ \forall w \in \Omega. \tag{31}$$

Applying the identity

$$(a - b)^T H(c - d) = \frac{1}{2}\{\|a - d\|_H^2 - \|a - c\|_H^2\} + \frac{1}{2}\{\|c - b\|_H^2 - \|d - b\|_H^2\},$$

to the right-hand side of (31) with

$$a = v, \quad b = \tilde{v}^k, \quad c = v^k, \quad \text{and} \quad d = v^{k+1},$$

we obtain

$$(v - \tilde{v}^k)^T H(v^k - v^{k+1}) = \frac{1}{2}(\|v - v^{k+1}\|_H^2 - \|v - v^k\|_H^2) + \frac{1}{2}(\|v^k - \tilde{v}^k\|_H^2 - \|v^{k+1} - \tilde{v}^k\|_H^2). \tag{32}$$

For the last term of (32), we have

$$\begin{aligned}
&\|v^k - \tilde{v}^k\|_H^2 - \|v^{k+1} - \tilde{v}^k\|_H^2 \\
&= \quad \|v^k - \tilde{v}^k\|_H^2 - \|(v^k - \tilde{v}^k) - (v^k - v^{k+1})\|_H^2 \\
&\overset{4.6a}{=} \quad \|v^k - \tilde{v}^k\|_H^2 - \|(v^k - \tilde{v}^k) - M(v^k - \tilde{v}^k)\|_H^2 \\
&= \quad 2(v^k - \tilde{v}^k)^T HM(v^k - \tilde{v}^k) - (v^k - \tilde{v}^k)^T M^T HM(v^k - \tilde{v}^k) \\
&= \quad (v^k - \tilde{v}^k)^T (Q^T + Q - M^T HM)(v^k - \tilde{v}^k) \\
&\overset{4.10}{=} (v^k - \tilde{v}^k)^T G(v^k - \tilde{v}^k).
\end{aligned} \tag{33}$$

Substituting (33) into (32), we get

$$(v - \tilde{v}^k)^T H(v^k - v^{k+1}) = \frac{1}{2}\big(\|v - v^{k+1}\|_H^2 - \|v - v^k\|_H^2\big) \\ + \frac{1}{2}(v^k - \tilde{v}^k)^T G(v^k - \tilde{v}^k). \quad (34)$$

Recall that $(w - \tilde{w}^k)^T F(\tilde{w}^k) = (w - \tilde{w}^k)^T F(w)$ (see (15)). Using this fact, the assertion of this lemma follows from (31) and (34) directly. □

When $G$ given in (30) is positive definite, as shown in [11], it is relatively easier to use the assertion (29) to prove the global convergence and estimate its worst-case convergence rate in terms of iteration complexity, see, e.g., [7, 14] for details and [6] (Sections 4 and 5 therein) for a tutorial proof. For the matrix $G$ given in (30), since $HM = Q$ and $M^T HM = M^T Q$, we have

$$M^T HM = \begin{pmatrix} I & 0 & -B^T \\ 0 & I & -C^T \\ 0 & 0 & I_m \end{pmatrix} \begin{pmatrix} (1+\tau)B^T B & 0 & 0 \\ 0 & (1+\tau)C^T C & 0 \\ -B & -C & I_m \end{pmatrix} \\ = \begin{pmatrix} (2+\tau)B^T B & B^T C & -B^T \\ C^T B & (2+\tau)C^T C & -C^T \\ -B & -C & I_m \end{pmatrix}.$$

Then, using (4.5b) and the above equation, we have

$$G = (Q^T + Q) - M^T HM \\ = \begin{pmatrix} (2+2\tau)B^T B & 0 & -B^T \\ 0 & (2+2\tau)C^T C & -C^T \\ -B & -C & 2I_m \end{pmatrix} \\ - \begin{pmatrix} (2+\tau)B^T B & B^T C & -B^T \\ C^T B & (2+\tau)C^T C & -C^T \\ -B & -C & I_m \end{pmatrix} \\ = \begin{pmatrix} \tau B^T B & -B^T C & 0 \\ -C^T B & \tau C^T C & 0 \\ 0 & 0 & I_m \end{pmatrix}. \quad (35)$$

By using the notation $D_0$ (see (19)), the matrix $G$ can be rewritten as

$$G = \begin{pmatrix} D_0 & \begin{matrix} 0 \\ 0 \end{matrix} \\ 0 \; 0 & I \end{pmatrix}.$$

Obviously, the proximal matrix $D_0$ in (19) can be rewritten as

$$D_0 = (\tau - 1)\begin{pmatrix} B^T B & 0 \\ 0 & C^T C \end{pmatrix} + \begin{pmatrix} B^T \\ -C^T \end{pmatrix}(B, -C). \quad (36)$$

Therefore, for $\tau \in (\frac{1}{2}, 1)$, $G$ is positive indefinite because the matrix $D_0$ is not so. The positive indefiniteness of $G$ is indeed the main difficulty of proving the convergence of the scheme (1.11) with $\tau > 0.5$; and we need to look into the quadratic term $(v^k - \tilde{v}^k)^T G(v^k - \tilde{v}^k)$ more intensively.

## 5 Investigation of the Quadratic Term $(v^k - \tilde{v}^k)^T G(v^k - \tilde{v}^k)$

As mentioned, the key point of proving the convergence of the scheme (1.11) with $\tau > 0.5$ is to analyze the quadratic term $(v^k - \tilde{v}^k)^T G(v^k - \tilde{v}^k)$ which is not guaranteed to be positive. In this section, we focus on investigating this term and show that

$$(v^k - \tilde{v}^k)^T G(v^k - \tilde{v}^k) \geq \psi(v^k, v^{k+1}) - \psi(v^{k-1}, v^k) + \varphi(v^k, v^{k+1}), \tag{37}$$

where $\psi(\cdot,\cdot)$ and $\varphi(\cdot,\cdot)$ are both non-negative functions. The first two terms $\psi(v^k, v^{k+1}) - \psi(v^{k-1}, v^k)$ in the right-hand side of (37) can be manipulated consecutively between iterates and the last term $\varphi(v^k, v^{k+1})$ should be such an error bound that can measure how much $w^{k+1}$ fails to be a solution point of (14). If we find such functions that guarantee the assertion (37), then we can substitute it into (29) and get the inequality

$$\begin{aligned}
&\theta(u) - \theta(\tilde{u}^k) + (w - \tilde{w}^k)^T F(w) \\
&\geq \frac{1}{2}\big(\|v - v^{k+1}\|_H^2 + \psi(v^k, v^{k+1})\big) - \frac{1}{2}\big(\|v - v^k\|_H^2 + \psi(v^{k-1}, v^k)\big) \\
&\quad + \frac{1}{2}\varphi(v^k, v^{k+1}), \quad \forall w \in \Omega.
\end{aligned} \tag{38}$$

As we shall show, all the components of the right-hand side of (38) in parentheses should be positive to establish the convergence and convergence rate of (1.11). It is indeed this requirement that implies our restriction of $\tau > 0.5$. We show the details in Theorem 5.5, preceded by several lemmas. Similar techniques for the convergence analysis of the ADMM are referred to, e.g. [4, 9, 10, 12].

**Lemma 5.1** *Let $\{w^k\}$ be the sequence generated by (1.11) for the problem (1) and $\tilde{w}^k$ be defined by (22). Then we have*

$$\begin{aligned}
&(v^k - \tilde{v}^k)^T G(v^k - \tilde{v}^k) \\
&= (1+\tau)\|B(y^k - y^{k+1})\|^2 + (1+\tau)\|C(z^k - z^{k+1})\|^2 + \|\lambda^k - \lambda^{k+1}\|^2 \\
&\quad + 2(\lambda^k - \lambda^{k+1})^T\big(B(y^k - y^{k+1}) + C(z^k - z^{k+1})\big).
\end{aligned} \tag{39}$$

***Proof*** First, according to (35), we have

$$G = \begin{pmatrix} \tau B^T B & -B^T C & 0 \\ -C^T B & \tau C^T C & 0 \\ 0 & 0 & I_m \end{pmatrix} = \begin{pmatrix} (1+\tau)B^T B & 0 & 0 \\ 0 & (1+\tau)C^T C & 0 \\ 0 & 0 & I_m \end{pmatrix} - \begin{pmatrix} B^T B & B^T C & 0 \\ C^T B & C^T C & 0 \\ 0 & 0 & 0 \end{pmatrix}$$

and thus

$$(v^k - \tilde{v}^k)^T G (v^k - \tilde{v}^k) = (1+\tau)\|B(y^k - \tilde{y}^k)\|^2 + (1+\tau)\|C(z^k - \tilde{z}^k)\|^2 + \|\lambda^k - \tilde{\lambda}^k\|^2 \\ -\|B(y^k - \tilde{y}^k) + C(z^k - \tilde{z}^k)\|^2.$$

For the term $\|\lambda^k - \tilde{\lambda}^k\|^2$ in the right-hand side of the above equation, because $\tilde{x}^k = x^{k+1}$,

$$\lambda^k - \tilde{\lambda}^k = Ax^{k+1} + By^k + Cz^k - b \quad \text{and} \quad Ax^{k+1} + By^{k+1} + Cz^{k+1} - b = \lambda^k - \lambda^{k+1},$$

we have

$$\lambda^k - \tilde{\lambda}^k = B(y^k - y^{k+1}) + C(z^k - z^{k+1}) + (\lambda^k - \lambda^{k+1}).$$

Finally, by a manipulation, we get

$$\begin{aligned} &(v^k - \tilde{v}^k)^T G (v^k - \tilde{v}^k) \\ &= (1+\tau)\|B(y^k - y^{k+1})\|^2 + (1+\tau)\|C(z^k - z^{k+1})\|^2 \\ &\quad -\|B(y^k - y^{k+1}) + C(z^k - z^{k+1})\|^2 \\ &\quad +\|B(y^k - y^{k+1}) + C(z^k - z^{k+1}) + (\lambda^k - \lambda^{k+1})\|^2 \\ &= (1+\tau)\|B(y^k - y^{k+1})\|^2 + (1+\tau)\|C(z^k - z^{k+1})\|^2 + \|\lambda^k - \lambda^{k+1}\|^2 \\ &\quad +2(\lambda^k - \lambda^{k+1})^T\big(B(y^k - y^{k+1}) + C(z^k - z^{k+1})\big). \end{aligned}$$

The lemma is proved. □

For further analysis, we will divide the crossing term $2(\lambda^k - \lambda^{k+1})^T\big(B(y^k - y^{k+1}) + C(z^k - z^{k+1})\big)$ in the right-hand side of (39) into two parts and give their lower bounds by quadratic terms.

**Lemma 5.2** *Let $\{w^k\}$ be the sequence generated by (1.11) for the problem (1) and $\tilde{w}^k$ be defined by (22). Then we have*

$$\begin{aligned} &(\lambda^k - \lambda^{k+1})^T\big(B(y^k - y^{k+1}) + C(z^k - z^{k+1})\big) \\ &\quad \geq \big(\psi(v^k, v^{k+1}) - \psi(v^{k-1}, v^k)\big) - 2(1-\tau)\big(\|B(y^k - y^{k+1})\|^2 + \|C(z^k - z^{k+1})\|^2\big), \end{aligned} \tag{40}$$

*where*

$$\psi(v^k, v^{k+1}) = \frac{1}{2}\left(\left\| \begin{matrix} y^k - y^{k+1} \\ z^k - z^{k+1} \end{matrix} \right\|_D^2 + (1-\tau)\big(\|B(y^k - y^{k+1})\|^2 + \|C(z^k - z^{k+1})\|^2\big)\right) \tag{41}$$

*with*

$$D = \begin{pmatrix} B^T \\ -C^T \end{pmatrix} (B, -C). \tag{42}$$

***Proof*** Recall (18). It holds that

$$(y^{k+1}, z^{k+1}) \in \mathcal{Y} \times \mathcal{Z}, \ \begin{pmatrix} \theta_2(y) - \theta_2(y^{k+1}) \\ \theta_3(z) - \theta_3(z^{k+1}) \end{pmatrix} + \begin{pmatrix} y - y^{k+1} \\ z - z^{k+1} \end{pmatrix}^T \left\{ \begin{pmatrix} -B^T \\ -C^T \end{pmatrix} \lambda^{k+1} + D_0 \begin{pmatrix} y^{k+1} - y^k \\ z^{k+1} - z^k \end{pmatrix} \right\} \geq 0, \ \forall (y, z) \in \mathcal{Y} \times \mathcal{Z}. \tag{43}$$

Analogously, for the previous iteration, we have

$$(y^k, z^k) \in \mathcal{Y} \times \mathcal{Z}, \ \begin{pmatrix} \theta_2(y) - \theta_2(y^k) \\ \theta_3(z) - \theta_3(z^k) \end{pmatrix} + \begin{pmatrix} y - y^k \\ z - z^k \end{pmatrix}^T \left\{ \begin{pmatrix} -B^T \\ -C^T \end{pmatrix} \lambda^k + D_0 \begin{pmatrix} y^k - y^{k-1} \\ z^k - z^{k-1} \end{pmatrix} \right\} \geq 0, \ \forall (y, z) \in \mathcal{Y} \times \mathcal{Z}. \tag{44}$$

Setting $(y, z) = (y^k, z^k)$ and $(y, z) = (y^{k+1}, z^{k+1})$ in (43) and (44), respectively, and adding them, we get

$$\begin{pmatrix} y^k - y^{k+1} \\ z^k - z^{k+1} \end{pmatrix}^T \left\{ \begin{pmatrix} B^T \\ C^T \end{pmatrix} (\lambda^k - \lambda^{k+1}) + D_0 \left[ \begin{pmatrix} y^{k+1} - y^k \\ z^{k+1} - z^k \end{pmatrix} - \begin{pmatrix} y^k - y^{k-1} \\ z^k - z^{k-1} \end{pmatrix} \right] \right\} \geq 0.$$

Consequently, we have

$$\begin{aligned} &(\lambda^k - \lambda^{k+1})^T \big(B(y^k - y^{k+1}) + C(z^k - z^{k+1})\big) \\ &\geq \begin{pmatrix} y^k - y^{k+1} \\ z^k - z^{k+1} \end{pmatrix}^T D_0 \left[ \begin{pmatrix} y^k - y^{k+1} \\ z^k - z^{k+1} \end{pmatrix} - \begin{pmatrix} y^{k-1} - y^k \\ z^{k-1} - z^k \end{pmatrix} \right]. \end{aligned} \tag{45}$$

From (19) and (42) we get

$$D_0 = D - (1 - \tau) \begin{pmatrix} B^T B & 0 \\ 0 & C^T C \end{pmatrix}.$$

Thus, using Cauchy-Schwaez inequality, from (45) we obtain

$$
\begin{aligned}
&(\lambda^k - \lambda^{k+1})^T\big(B(y^k - y^{k+1}) + C(z^k - z^{k+1})\big)\\
&\geq \begin{pmatrix} y^k - y^{k+1} \\ z^k - z^{k+1} \end{pmatrix}^T \left\{ D - (1-\tau)\begin{pmatrix} B^T B & 0 \\ 0 & C^T C \end{pmatrix}\right\}\left[\begin{pmatrix} y^k - y^{k+1} \\ z^k - z^{k+1} \end{pmatrix} - \begin{pmatrix} y^{k-1} - y^k \\ z^{k-1} - z^k \end{pmatrix}\right]\\
&= \left\| \begin{matrix} y^k - y^{k+1} \\ z^k - z^{k+1} \end{matrix} \right\|_D^2 - \begin{pmatrix} y^k - y^{k+1} \\ z^k - z^{k+1} \end{pmatrix}^T D \begin{pmatrix} y^{k-1} - y^k \\ z^{k-1} - z^k \end{pmatrix}\\
&\quad -(1-\tau)\Big(\|B(y^k - y^{k+1})\|^2 + \|C(z^k - z^{k+1})\|^2\Big)\\
&\quad +(1-\tau)\begin{pmatrix} y^k - y^{k+1} \\ z^k - z^{k+1} \end{pmatrix}^T \begin{pmatrix} B^T B & 0 \\ 0 & C^T C \end{pmatrix}\begin{pmatrix} y^{k-1} - y^k \\ z^{k-1} - z^k \end{pmatrix}\\
&\geq \frac{1}{2}\left\| \begin{matrix} y^k - y^{k+1} \\ z^k - z^{k+1} \end{matrix} \right\|_D^2 - \frac{1}{2}\left\| \begin{matrix} y^{k-1} - y^k \\ z^{k-1} - z^k \end{matrix} \right\|_D^2\\
&\quad -\frac{3}{2}(1-\tau)\Big(\|B(y^k - y^{k+1})\|^2 + \|C(z^k - z^{k+1})\|^2\Big)\\
&\quad -\frac{1}{2}(1-\tau)\Big(\|B(y^{k-1} - y^k)\|^2 + \|C(z^{k-1} - z^k)\|^2\Big),
\end{aligned} \tag{46}
$$

where the last inequality is because of the Cauchy-Schwarz inequality. Manipulating the right-hand side of (46) recursively and using the notation of $\psi(\cdot,\cdot)$ (see (41)), we get (40) and the lemma is proved. □

In addition to (40), we need to the term $(\lambda^k - \lambda^{k+1})^T\big(B(y^k - y^{k+1}) + C(z^k - z^{k+1})\big)$ by an another quadratic terms. This is done by the following lemma.

**Lemma 5.3** *Let $\{w^k\}$ be the sequence generated by (1.11) for the problem (1) and $\tilde{w}^k$ be defined by (22). Then, for $\tau \in (0.5, 1)$, we have*

$$
\begin{aligned}
&(\lambda^k - \lambda^{k+1})^T\big(B(y^k - y^{k+1}) + C(z^k - z^{k+1})\big)\\
&\geq -\tau\Big(\|B(y^k - y^{k+1})\|^2 + \|C(z^k - z^{k+1})\|^2\Big) - \Big(\frac{3}{2} - \tau\Big)\|\lambda^k - \lambda^{k+1}\|^2.
\end{aligned} \tag{47}
$$

***Proof*** Setting $\delta = \tau - \frac{1}{2}$. Because $\tau \in (0.5, 1)$, we have $\delta \in (0, 0.5)$. Using the Cauchy-Schwarz inequality twice, we get

$$
\begin{aligned}
&(\lambda^k - \lambda^{k+1})^T\big(B(y^k - y^{k+1}) + C(z^k - z^{k+1})\big)\\
&\geq -\frac{1}{4(1-\delta)}\|B(y^k - y^{k+1}) + C(z^k - z^{k+1})\|^2 - (1-\delta)\|\lambda^k - \lambda^{k+1}\|^2\\
&\geq -\frac{1}{2(1-\delta)}\big(\|B(y^k - y^{k+1})\|^2 + \|C(z^k - z^{k+1})\|^2\big) - (1-\delta)\|\lambda^k - \lambda^{k+1}\|^2.
\end{aligned}
$$

Since $\delta \in (0, 0.5)$, we have

$$
\frac{1}{2(1-\delta)} < \frac{1}{2} + \delta,
$$

and thus

$$
\begin{aligned}
&(\lambda^k - \lambda^{k+1})^T \big(B(y^k - y^{k+1}) + C(z^k - z^{k+1})\big) \\
&\quad \geq -\big(\frac{1}{2} + \delta\big)\big(\|B(y^k - y^{k+1})\|^2 + \|C(z^k - z^{k+1})\|^2\big) - (1-\delta)\|\lambda^k - \lambda^{k+1}\|^2.
\end{aligned}
$$

Substituting $\delta = \tau - \frac{1}{2}$ in the above inequality, we get (47) and the lemma is proved. □

Recall that we want to bound the quadratic term $(v^k - \tilde{v}^k)^T G(v^k - \tilde{v}^k)$ in the form of (38). Our previous analysis enables us to achieve it; and this is the basis of the convergence analysis to be shown soon.

**Lemma 5.4** *Let* $\{w^k\}$ *be the sequence generated by (1.11) for the problem (1) and* $\tilde{w}^k$ *be defined by (22). Then, for* $\tau \in (0.5, 1)$, *we have*

$$
(v^k - \tilde{v}^k)^T G(v^k - \tilde{v}^k) \geq \big(\psi(v^k, v^{k+1}) - \psi(v^{k-1}, v^k)\big) + \varphi(v^k, v^{k+1}), \tag{48}
$$

*where* $\psi(v^k, v^{k+1})$ *is defined in (41) and*

$$
\varphi(v^k, v^{k+1}) = \big(\tau - \frac{1}{2}\big)\big(2\|B(y^k - y^{k+1})\|^2 + 2\|C(z^k - z^{k+1})\|^2 + \|\lambda^k - \lambda^{k+1}\|^2\big). \tag{49}
$$

***Proof*** Substituting (40) and (47) into (39), we get

$$
\begin{aligned}
&(v^k - \tilde{v}^k)^T G(v^k - \tilde{v}^k) \\
&\geq (1+\tau)\|B(y^k - y^{k+1})\|^2 + (1+\tau)\|C(z^k - z^{k+1})\|^2 + \|\lambda^k - \lambda^{k+1}\|^2 \\
&\quad + \big(\psi(v^k, v^{k+1}) - \psi(v^{k-1}, v^k)\big) - 2(1-\tau)\big(\|B(y^k - y^{k+1})\|^2 + \|C(z^k - z^{k+1})\|^2\big) \\
&\quad - \tau\big(\|B(y^k - y^{k+1})\|^2 + \|C(z^k - z^{k+1})\|^2\big) - \big(\frac{3}{2} - \tau\big)\|\lambda^k - \lambda^{k+1}\|^2 \\
&= \big(\psi(v^k, v^{k+1}) - \psi(v^{k-1}, v^k)\big) \\
&\quad + (2\tau - 1)\big(\|B(y^k - y^{k+1})\|^2 + \|C(z^k - z^{k+1})\|^2\big) + \big(\tau - \frac{1}{2}\big)\|\lambda^k - \lambda^{k+1}\|^2.
\end{aligned}
$$

The assertion of this lemma follows from the definition of $\varphi(v^k, v^{k+1})$ directly.□

Finally, substituting (48) into (29), we obtain the following theorem directly. This theorem plays a fundamental role in proving the convergence of (1.11) with $\tau > 0.5$.

**Theorem 5.5** *Let $\{w^k\}$ be the sequence generated by (1.11) for the problem (1) and $\tilde{w}^k$ be defined by (22). Then we have*

$$\theta(u) - \theta(\tilde{u}^k) + (w - \tilde{w}^k)^T F(w) \geq \frac{1}{2}\big(\|v - v^{k+1}\|_H^2 + \psi(v^k, v^{k+1})\big) - \frac{1}{2}\big(\|v - v^k\|_H^2 + \psi(v^{k-1}, v^k)\big) + \frac{1}{2}\varphi(v^k, v^{k+1}), \quad \forall w \in \Omega, \tag{50}$$

*where $\psi(v^k, v^{k+1})$ and $\varphi(v^k, v^{k+1})$ are defined in (41) and (49), respectively.*

## 6 Convergence

As mentioned, proving the convergence of the scheme (1.11) with $\tau > 0.5$ essentially relies on Theorem 5.5. With Theorem 5.5, the remaining part of the proof is subroutine. In this section, we present the convergence of the scheme (1.11) with $\tau > 0.5$; a lemma is first proved to show the contraction property of the sequence generated by (1.11).

**Lemma 6.1** *Let $\{w^k\}$ be the sequence generated by (1.11) with $\tau > 0.5$ for the problem (1). Then we have*

$$\big(\|v^{k+1} - v^*\|_H^2 + \psi(v^k, v^{k+1})\big) \leq \big(\|v^k - v^*\|_H^2 + \psi(v^{k-1}, v^k)\big) - \varphi(v^k, v^{k+1}), \tag{51}$$

*where $\psi(v^k, v^{k+1})$ and $\varphi(v^k, v^{k+1})$ are defined in (41) and (49), respectively.*

***Proof*** Setting $w = w^*$ in (50) and using

$$\theta(\tilde{u}^k) - \theta(u^*) + (\tilde{w}^k - w^*)^T F(w^*) \geq 0,$$

we obtain the assertion (51) immediately. □

**Theorem 6.2** *Let $\{w^k\}$ be the sequence generated by (1.11) with $\tau > 0.5$ for the problem (1). Then the sequence $\{v^k\}$ converges to a $v^\infty \in \mathcal{V}^*$ when $B$ and $C$ are both full column rank.*

***Proof*** First, it follows from (51) and (49) that

$$(\tau - \frac{1}{2})\big(2\|B(y^k - y^{k+1})\|^2 + 2\|C(z^k - z^{k+1})\|^2 + \|\lambda^k - \lambda^{k+1}\|^2\big)$$
$$\leq \big(\|v^k - v^*\|_H^2 + \psi(v^{k-1}, v^k)\big) - \big(\|v^{k+1} - v^*\|_H^2 + \psi(v^k, v^{k+1})\big).$$

Summarizing the last inequality over $k = 1, 2, \ldots$, we obtain

$$\sum_{k=1}^{\infty}\Big\{(\tau-\frac{1}{2})\Big(2\|B(y^k-y^{k+1})\|^2+2\|C(z^k-z^{k+1})\|^2+\|\lambda^k-\lambda^{k+1}\|^2\Big)\Big\}$$
$$\leq \|v^1-v^*\|_H^2+\psi(v^0,v^1)$$

and thus

$$\lim_{k\to\infty}\|B(y^k-y^{k+1})\|^2+\|C(z^k-z^{k+1})\|^2+\|\lambda^k-\lambda^{k+1}\|^2=0. \tag{52}$$

For an arbitrarily fixed $v^*\in\mathcal{V}^*$, it follows from (51) that, for any $k>1$, we have

$$\|v^{k+1}-v^*\|_H^2 \leq \|v^k-v^*\|_H^2+\psi(v^{k-1},v^k) \leq \|v^1-v^*\|_H^2+\psi(v^0,v^1). \tag{53}$$

Thus the sequence $\{v^k\}$ is bounded. Because $M$ is non-singular, according to (4.6), $\{\tilde{v}^k\}$ is also bounded. Let $v^\infty$ be a cluster point $\{\tilde{v}^k\}$ and $\{\tilde{v}^{k_j}\}$ be the subsequence of $\{\tilde{v}^k\}$ converging to $v^\infty$. Let $x^\infty$ be the vector induced by given $(y^\infty,z^\infty,\lambda^\infty)\in\mathcal{V}$. Then, it follows from (31) that

$$w^\infty\in\Omega,\quad \theta(u)-\theta(u^\infty)+(w-w^\infty)^TF(w^\infty)\geq 0,\quad \forall w\in\Omega,$$

which means $w^\infty$ is a solution point of (14) and its essential part $v^\infty\in\mathcal{V}^*$. Since $v^\infty\in\mathcal{V}^*$, it follows from (53) that

$$\|v^{k+1}-v^\infty\|_H^2\leq\|v^k-v^\infty\|_H^2+\psi(v^{k-1},v^k). \tag{54}$$

Together with (52), it is impossible that the sequence $\{v^k\}$ has more than one cluster point. Thus $\{v^k\}$ converges to $v^\infty$ and the proof is complete. □

**Remark 6.3** Note that the convergence of (1.11) with $\tau>0.5$ in terms of the sequence $\{v^k\}$ is proved in Theorem 6.2 under the assumption that both $B$ and $C$ are full column rank. Without this assumption, weaker convergence results in terms of $\{By^k,Cz^k\}$ can be derived. We refer to Sect. 6 in [11] for details.

## 7 The Optimality of $\tau=0.5$

We have proved the convergence of (1.11) with $\tau>0.5$; the key is sufficiently ensuring the non-negativeness of the coefficients in the right-hand side of (50). In this section, we show by an example that any $\tau\in(0,0.5)$ may yield divergence of (1.11). Hence, $\tau=0.5$ is the watershed, or optimal value, to ensure the convergence of (1.11).

For any given $\tau < 0.5$, we take $\epsilon = 0.5 - \tau > 0$ and consider the problem

$$\min\{x + \frac{\epsilon}{2}y^2 + \frac{\epsilon}{2}z^2 \mid x + y + z = 0,\ x \in \{0\}, y \in \Re, z \in \Re\}, \tag{55}$$

which is a special case of the model (1). Obviously, the solution of this problem is $x = y = z = 0$.

The augmented Lagrangian function of the problem (55) with a penalty parameter of 1 is

$$\mathcal{L}(x, y, z, \lambda) = x + \frac{\epsilon}{2}y^2 + \frac{\epsilon}{2}z^2 - \lambda^T(x + y + z) + \frac{1}{2}\|x + y + z\|^2;$$

and the iterative scheme (1.11) for (55) is

$$\begin{cases} x^{k+1} = \arg\min\{\mathcal{L}(x, y^k, z^k, \lambda^k) \mid x \in \{0\}\}, \\ y^{k+1} = \arg\min\{\mathcal{L}(x^{k+1}, y, z^k, \lambda^k) + \frac{\tau}{2}\|y - y^k\|^2 \mid y \in \Re\}, \\ z^{k+1} = \arg\min\{\mathcal{L}(x^{k+1}, y^k, z, \lambda^k) + \frac{\tau}{2}\|z - z^k\|^2 \mid z \in \Re\}, \\ \lambda^{k+1} = \lambda^k - (x^{k+1} + y^{k+1} + z^{k+1}). \end{cases} \tag{56}$$

Since $\mathcal{X} = \{0\}$, we have $x^{k+1} \equiv 0$. Ignoring constant terms in the objective function of the subproblems, the recursion (56) becomes

$$\begin{cases} x^{k+1} \equiv 0, \\ y^{k+1} = \arg\min\{\frac{\epsilon}{2}y^2 - y^T\lambda^k + \frac{1}{2}\|y + z^k\|^2 + \frac{\tau}{2}\|y - y^k\|^2 \mid y \in \Re\}, \\ z^{k+1} = \arg\min\{\frac{\epsilon}{2}z^2 - z^T\lambda^k + \frac{1}{2}\|y^k + z\|^2 + \frac{\tau}{2}\|z - z^k\|^2 \mid z \in \Re\}, \\ \lambda^{k+1} = \lambda^k - (y^{k+1} + z^{k+1}). \end{cases} \tag{57}$$

Further, it follows from (57) that

$$\begin{cases} \epsilon y^{k+1} - \lambda^k + (y^{k+1} + z^k) + \tau(y^{k+1} - y^k) = 0, \\ \epsilon z^{k+1} - \lambda^k + (z^{k+1} + y^k) + \tau(z^{k+1} - z^k) = 0, \\ \lambda^{k+1} = \lambda^k - (y^{k+1} + z^{k+1}). \end{cases}$$

Thus, the iterative scheme for $v = (y, z, \lambda)$ can be written as

$$\begin{cases} (\tau + 1 + \epsilon)y^{k+1} = \tau y^k - z^k + \lambda^k, \\ (\tau + 1 + \epsilon)z^{k+1} = -y^k + \tau z^k + \lambda^k, \\ \lambda^{k+1} = \lambda^k - (y^{k+1} + z^{k+1}). \end{cases} \tag{58}$$

Without loss of generality, we can take $y^0 = z^0$ and thus $y^k \equiv z^k$, for all $k > 0$. Using this fact and $\tau + \epsilon = 0.5$, we get

$$\begin{cases} \frac{3}{2}y^{k+1} = (\tau - 1)y^k + \lambda^k, \\ \lambda^{k+1} = \lambda^k - 2y^{k+1}. \end{cases} \tag{59}$$

With elementary manipulations, we obtain

$$\begin{cases} y^{k+1} = \frac{-2(1-\tau)}{3} y^k + \frac{2}{3}\lambda^k, \\ \lambda^{k+1} = \frac{4(1-\tau)}{3} y^k + \frac{-1}{3}\lambda^k, \end{cases} \tag{60}$$

which can be written as

$$\begin{pmatrix} y^{k+1} \\ \lambda^{k+1} \end{pmatrix} = P(\tau) \begin{pmatrix} y^k \\ \lambda^k \end{pmatrix} \quad \text{with} \quad P(\tau) = \frac{1}{3} \begin{pmatrix} -2(1-\tau) & 2 \\ 4(1-\tau) & -1 \end{pmatrix}. \tag{61}$$

Let $f_1(\tau)$ and $f_2(\tau)$ be the two eigenvalues of the matrix $P(\tau)$. Then we have

$$f_1(\tau) = \frac{1}{6}\Big((2\tau - 3) + \sqrt{(3-2\tau)^2 + 24(1-\tau)}\Big),$$

and

$$f_2(\tau) = \frac{1}{6}\Big((2\tau - 3) - \sqrt{(3-2\tau)^2 + 24(1-\tau)}\Big).$$

Certainly, the scheme (60) is divergent if the absolute value of one of the eigenvalues of the matrix $P(\tau)$ is greater than 1. Indeed, it holds that $f_2(\tau) < -1$ for any $\tau \in (0, 0.5)$. To see this assertion, we notice that

$$\begin{aligned} f_2(\tau) < -1 \Leftrightarrow (2\tau - 3) - \sqrt{(3-2\tau)^2 + 24(1-\tau)} &< -6 \\ \Leftrightarrow \qquad 2\tau + 3 &< \sqrt{4\tau^2 - 36\tau + 33} \\ \Leftrightarrow \qquad 4\tau^2 + 12\tau + 9 &< 4\tau^2 - 36\tau + 33 \\ \Leftrightarrow \qquad \tau &< 0.5. \end{aligned}$$

Hence, the scheme (1.11) is not necessarily convergent for any $\tau \in (0, 0.5)$.

## 8 Convergence Rate

In this section, we derive a worst-case $O(1/t)$ convergence rate in terms of iteration complexity for the scheme (1.11) with $\tau > 0.5$, where $t$ is the iteration counter. Hence, although the condition $\tau \geq 1$ in [11] is now relaxed to $\tau > 0.5$, the same convergence rate result in [11] remains valid for the scheme (1.11). Similar analysis is refereed to [11, 13].

First of all, recall (14). If we find $\tilde{w}$ satisfying the inequality

$$\tilde{w} \in \Omega, \quad \theta(u) - \theta(\tilde{u}) + (w - \tilde{w})^T F(\tilde{w}) \geq 0, \quad \forall\, w \in \Omega,$$

then $\tilde{w}$ is a solution point of (14). As mentioned in (15), we have $(w-\tilde{w})^T F(\tilde{w}) = (w-\tilde{w})^T F(\tilde{w})$. Thus, a solution point $\tilde{w}$ of (14) can be also characterized by

$$\tilde{w} \in \Omega, \quad \theta(u) - \theta(\tilde{u}) + (w-\tilde{w})^T F(w) \geq 0, \quad \forall\, w \in \Omega.$$

Therefore, as [3], for given $\epsilon > 0$, $\tilde{w} \in \Omega$ is called an $\epsilon$-approximate solution of VI$(\Omega, F, \theta)$ if it satisfies

$$\tilde{w} \in \Omega, \quad \theta(u) - \theta(\tilde{u}) + (w-\tilde{w})^T F(w) \geq -\epsilon, \quad \forall\, w \in \mathcal{D}_{(\tilde{w})},$$

where

$$\mathcal{D}_{(\tilde{w})} = \{w \in \Omega \mid \|w - \tilde{w}\| \leq 1\}.$$

In the following, we show that based on the first $t$ iterates generated by the scheme (1.11) with $\tau > 0.5$, we can find an approximate solution of (14), denoted by $\tilde{w} \in \Omega$, such that

$$\tilde{w} \in \Omega \quad \text{and} \quad \sup_{w \in \mathcal{D}_{(\tilde{w})}} \left\{\theta(\tilde{u}) - \theta(u) + (\tilde{w} - w)^T F(w)\right\} \leq \epsilon, \tag{62}$$

where $\epsilon = O(1/t)$. That is, a worst-case $O(1/t)$ convergence rate is established for the scheme (1.11) with $\tau > 0.5$. Theorem 5.5 is still the basis for the analysis in this section.

**Theorem 8.1** *Let $\{w^k\}$ be the sequence generated by (1.11) with $\tau > 0.5$ for the problem (1) and $\tilde{w}^k$ be defined by (22). Then for any integer t, we have*

$$\theta(\tilde{u}_t) - \theta(u) + (\tilde{w}_t - w)^T F(w) \leq \frac{1}{2t}\left\{\|v - v^1\|_H^2 + \psi(v^0, v^1)\right\}, \tag{63}$$

*where*

$$\tilde{w}_t = \frac{1}{t}\Big(\sum_{k=1}^{t} \tilde{w}^k\Big) \tag{64}$$

*and $\psi(v^0, v^1)$ is defined in (41) and thus*

$$\psi(v^0, v^1) = \frac{1}{2}\left(\left\|\begin{matrix} y^0 - y^1 \\ z^0 - z^1 \end{matrix}\right\|_D^2 + (1-\tau)\Big(\|B(y^0 - y^1)\|^2 + \|C(z^0 - z^1)\|^2\Big)\right).$$

***Proof*** First, it follows from (50) that

$$\theta(u) - \theta(\tilde{u}^k) + (w - \tilde{w}^k)^T F(w) \geq \frac{1}{2}\Big(\|v - v^{k+1}\|_H^2 + \psi(v^k, v^{k+1})\Big) - \frac{1}{2}\Big(\|v - v^k\|_H^2 + \psi(v^{k-1}, v^k)\Big).$$

Thus, we have

$$\theta(\tilde{u}^k) - \theta(u) + (\tilde{w}^k - w)^T F(w) + \frac{1}{2}\big(\|v - v^{k+1}\|_H^2 + \psi(v^k, v^{k+1})\big) \le \frac{1}{2}\big(\|v - v^k\|_H^2 + \psi(v^{k-1}, v^k)\big). \tag{65}$$

Summarizing the inequality (65) over $k = 1, 2, \dots, t$, we obtain

$$\sum_{k=1}^{t} \theta(\tilde{u}^k) - t\theta(u) + (\sum_{k=1}^{t} \tilde{w}^k - tw)^T F(w) \le \frac{1}{2}\big(\|v - v^1\|_H^2 + \psi(v^0, v^1)\big)$$

and thus

$$\frac{1}{t}\big(\sum_{k=1}^{t} \theta(\tilde{u}^k)\big) - \theta(u) + (\tilde{w}_t - w)^T F(w) \le \frac{1}{2t}\big(\|v - v^1\|_H^2 + \psi(v^0, v^1)\big). \tag{66}$$

Since $\theta(u)$ is convex and

$$\tilde{u}_t = \frac{1}{t}\big(\sum_{k=1}^{t} \tilde{u}^k\big),$$

we have that

$$\theta(\tilde{u}_t) \le \frac{1}{t}\big(\sum_{k=1}^{t} \theta(\tilde{u}^k)\big).$$

Substituting it into (66), the assertion of this theorem follows directly. □

For a given compact set $\mathcal{D}_{(\tilde{w})} \subset \Omega$, let

$$d := \sup\left\{ \|v - v^1\|_H^2 + \frac{1}{2}\left\| \begin{matrix} y^0 - y^1 \\ z^0 - z^1 \end{matrix} \right\|_D^2 + \frac{1-\tau}{2}\big(\|B(y^0 - y^1)\|^2 + \|C(z^0 - z^1)\|^2\big) \,\Big|\, w \in \mathcal{D}_{(\tilde{w})} \right\}$$

where $v^0 = (y^0, z^0, \lambda^0)$ and $v^1 = (y^1, z^1, \lambda^1)$ are the initial and the first generated iterates, respectively. Then, after $t$ iterations of the scheme (1.11), the point $\tilde{w}_t \in \Omega$ defined in (64) satisfies

$$\tilde{w} \in \Omega \quad \text{and} \quad \sup_{w \in \mathcal{D}_{(\tilde{w})}} \big\{\theta(\tilde{u}) - \theta(u) + (\tilde{w} - w)^T F(w)\big\} \le \frac{d}{2t} = O(\frac{1}{t}),$$

which means $\tilde{w}_t$ is an approximate solution of VI$(\Omega, F, \theta)$ with an accuracy $O(1/t)$ (recall (62)). That is, a worst-case $O(1/t)$ convergence rate is established for the scheme (1.11) with $\tau > 0.5$. Since $\tilde{w}_t$ defined in (64) is the average of all iterates of (1.11), this convergence rate is in the ergodic sense.

## 9 Conclusions

We revisit the splitting method proposed in [11] for solving separable convex minimization models; and show that its optimal proximal parameter is 0.5 when the objective function is the sum of three functions. This optimal proximal parameter offers the possibility of immediate numerical acceleration; which can be easily verified by the examples tested in [2, 11] and others. For succinctness, we omit the presentation of numerical results. Meanwhile, more sophisticated techniques are required for the convergence analysis because this optimal proximal parameter generates positive indefiniteness in the proximal regularization term as well. We establish the convergence and estimate the worst-case convergence rate in terms of iteration complexity for the improved version of the method in [11] with the optimal proximal parameter. This work is inspired by the analysis in our recent work [9, 10] for the augmented Lagrangian method and alternating direction method of multipliers.

## References


1. C.H. Chen, B.S. He, Y.Y. Ye, X.M. Yuan, The direct extension of ADMM for multi-block convex minimization problems is not necessary convergent. Math. Program. **155**, 57–79 (2016)
2. E. Esser, M. Möller, S. Osher, G. Sapiro, J. Xin, A convex model for non-negative matrix factorization and dimensionality reduction on physical space. IEEE Trans. Imaging Process. **21**(7), 3239–3252 (2012)
3. F. Facchinei, J.S. Pang, *Finite-Dimensional Variational Inequalities and Complementarity Problems*, Vol. I (Springer Series in Operations Research, Springer, 2003)
4. R. Glowinski, *Numerical Methods for Nonlinear Variational Problems* (Springer, New York, Berlin, Heidelberg, Tokyo, 1984)
5. R. Glowinski, A. Marrocco, *Sur l'approximationparéléments finis d'ordre un et larésolution parpénalisation-dualité d'une classe de problèmes deDirichlet non linéaires*, Revue Fr. Autom. Inform. Rech.Opér., Anal. Numér. 2 (1975), pp. 41–76
6. B.S. He, PPA-like contraction methods for convex optimization: a framework using variational inequality approach. J. Oper. Res. Soc. China **3**, 391–420 (2015)
7. B.S. He, H. Liu, Z.R. Wang, X.M. Yuan, A strictly contractive Peaceman-Rachford splitting method for convex programming. SIAM J. Optim. **24**, 1011–1040 (2014)
8. B.S. He, L.S. Hou, X.M. Yuan, On full Jacobian decomposition of the augmented Lagrangian method for separable convex programming. SIAM J. Optim. **25**(4), 2274–2312 (2015)
9. B.S. He, F. Ma, X.M. Yuan, Indefinite proximal augmented Lagrangian method and its application to full Jacobian splitting for multi-block separable convex minimization problems. IMA J. Numer. Anal. **75**, 361–388 (2020)
10. B.S. He, F. Ma, X.M. Yuan, Optimally linearizing the alternating direction method of multipliers for convex programming. Comput. Optim. Appl. **75**(2), 361–388 (2020)
11. B.S. He, M. Tao, X.M. Yuan, A splitting method for separable convex programming. IMA J. Numer. Anal. **35**, 394–426 (2014)
12. B.S. He, H. Yang, Some convergence properties of a method of multipliers for linearly constrained monotone variational inequalities. Oper. Res. Lett. **23**, 151–161 (1998)
13. B.S. He, X.M. Yuan, On the $O(1/t)$ convergence rate of the alternating direction method. SIAM J. Numer. Anal. **50**, 700–709 (2012)
14. B.S. He, X.M. Yuan, On non-ergodic convergence rate of Douglas-Rachford alternating directions method of multipliers. Numer. Math. **130**, 567–577 (2015)

15. M.R. Hestenes, Multiplier and gradient methods. J. Optim. Theory Appl. **4**, 303–320 (1969)
16. K.C. Kiwiel, C.H. Rosa, A. Ruszczyński, Proximal decomposition via alternating linearization. SIAM J. Optim. **9**, 668–C689 (1999)
17. B. Martinet, Regularisation, d'inéquations variationelles par approximations succesives. Rev. Francaise d'Inform. Recherche Oper. **4**, 154–159 (1970)
18. Powell M.J.D., A method for nonlinear constraints in minimization problems, in *Optimization*, ed. by R. Fletcher (Academic Press, New York, NY, 1969), pp. 283–298
19. R.T. Rockafellar, Monotone operators and the proximal point algorithm. SIAM J. Cont. Optim. **14**, 877–898 (1976)
20. M. Tao, X.M. Yuan, Recovering low-rank and sparse components of matrices from incomplete and noisy observations. SIAM J. Optim. **21**, 57–81 (2011)
21. R. Tibshirani, M. Saunders, S. Rosset, J. Zhu, K. Knight, Sparsity and smoothness via the fused lasso. J. R. Stat. Soc. **67**, 91–108 (2005)
22. X. Zhou, C. Yang, W. Yu, Moving object detection by detecting contiguous outliers in the low-rank representation. IEEE Trans. Pattern Anal. Mach. Intell. **35**, 597–610 (2013)
23. Z. Zhou, X. Li, J. Wright, E.J. Candes, Y. Ma, Stable principal component pursuit, in *Proceedings of international symposium on information theory*, Austin, Texas, USA (2010)

# A New Initialization Method for Neural Networks with Weight Sharing

**Xiaofeng Ding, Hongfei Yang, Raymond H. Chan, Hui Hu, Yaxin Peng, and Tieyong Zeng**

**Abstract** A proper initialization of parameters in a neural network can facilitate its training. The Xavier initialization introduced by Glorot and Bengio which is later generalized to Kaiming initialization by He, Zhang, Ren and Sun are now widely used. However, from experiments we find that networks with heavy weight sharing are difficulty to train even with the Xavier or the Kaiming initialization. We also notice that a certain simple network can be decomposed in two ways, where one is difficult to train while the other is easy to train, when both are properly initialized by the Xavier or the Kaiming initialization. In this paper we will propose a new initialization

Raymond H. Chan—The work was supported by the HKRGC Grant No. CUHK14301718. Hui Hu—The work was supported by the National Natural Science Foundation of China under Grant 11771276.Correspondence author. Tieyong Zeng—The work was supported by the NSFC under Grant 11671002, in part by the CUHK Start-Up, and in part by the CUHK DAG under Grant 4053342, Grant 4053405, Grant RGC 14300219, RGC 14302920, and Grant NSFC/RGC N_CUHK 415/19.

X. Ding · Y. Peng
Department of Mathematics, Shanghai University, Shanghai, China
e-mail: dxfeng@shu.edu.cn

Y. Peng
e-mail: yaxin.peng@shu.edu.cn

H. Yang · R. H. Chan
Department of Mathematics, City University of Hong Kong, Hong Kong, China
e-mail: honyang@cityu.edu.hk

R. H. Chan
e-mail: rchan.sci@cityu.edu.hk

H. Hu
HiSilicon Technologies Co., Limited, Shenzhen, China
e-mail: huhui12@huawei.com

T. Zeng (✉)
Department of Mathematics, the Chinese University of Hong Kong, Hong Kong, China
e-mail: zeng@math.cuhk.edu.hk



X.-C. Tai et al. (eds.), *Mathematical Methods in Image Processing and Inverse Problems*, Springer Proceedings in Mathematics & Statistics 360,
https://doi.org/10.1007/978-981-16-2701-9_9

method which will increase training speed and training stability of neural networks with heavy weight sharing. We will also propose a simple yet efficient method to adjust learning rates layer by layer which is indispensable to our initialization.

**Keywords** Learning rate · Neural networks · Weight sharing · Xavier initialization

## 1 Introduction

In recent years deep learning methods have achieved remarkable progresses in a broad range of tasks including object classifications [10, 15], semantic segmentation [5, 19], natural language processing [2, 3], and speech recognition [1, 12]. In 2016, AlphaGo, a Go game software powered by deep learning algorithms [22], became the first Go game software to beat a human champion in a series of 5 Go games.

How to effectively train deep neural networks is an active research area. In the seminal work of Xavier and Yoshua [6] they observed that a proper initialization of the learnable parameters played an important role in training a neural network. They proposed a novel initialization method for layers with activation functions that can be approximated by linear functions. Later He et al. in [11] extended the method of Xavier and Yoshua to consider highly non-linear activation functions like the ReLU or the leaky-ReLU activation functions. Nowadays these two initialization methods are widely used in deep learning software packages like TensorFlow, PyTorch and Keras. Since these two initialization methods only differ by a scalar factor called "gain", which is decided by the activation function, in this paper we regard them as one method and call it the Xavier/Kaiming initialization.

The main idea of the Xavier/Kaiming initialization is to maintain a constant variance for different layers in both the forward and the backward propagations. However, we notice that when there is heavy weigh sharing for certain parameters, the activation variances and the variance of the back-propagated gradients are not good indicators of variances of learnable parameters. As a result, the updates of heavily shared parameters may be much faster than other parameters, and this leads to difficulties in training. We also noticed that if we multiply one layer by a positive constant number and then divide this number in another layer, we can define a network which is very difficult to train even if we use the Xavier/Kaiming initialization. These observations lead us to propose a new initialization.

There are some other initialization methods. In [21] the authors proposed a random orthogonal initialization condition based on their analysis of dynamics of deep learning, and this initialization method is widely used in networks with recurrent structures [13, 17]. This random orthogonal initial condition is further generalized in [18]. There are also initialization methods based on the mean field theory [9, 24, 25].

In this paper, we will propose a new initialization method based on the Xavier/Kaiming initialization. For a fully connected layer with no weight sharing, our new method is the same as the Xavier/Kaiming initialization method. However, our method can cope with weight sharing, which is a common phenomenon for CirCNN implementation of neural networks (described below). We will give various numerical examples to verify the effectiveness of our initialization method.

Another contribution of this paper is that we introduce a simple yet efficient way to adjust learning rates of parameters layer by layer (see (13)). The main idea is to multiply the weight matrix of each layer by a positive constant scalar, and at the same time change the initialization of layers accordingly. Actually this scalar-multiplying technique is indispensable to our proposed initialization method which will be demonstrated below.

This paper is organized as the following. In the next section, we briefly give the notations and terms used in this paper. In the third section, we introduce the Xavier/Kaiming initialization and its limitation. In the forth section, we propose our new method for the fully connected case, and we will show how our new method can overcome the limitation. Lastly we conclude with possible future research.

## 2 Neural Networks and CirCNN Implementations

In this section we give the notations and terms used in this paper. We will also introduce CirCNN networks which will be used in our numerical experiments.

### 2.1 Single Layer Structure

A typical layer in a multi-layer neural network has the form

$$y = \mathrm{act}(Wx + b), \tag{1}$$

where $x \in \mathbb{R}^M$ is the input vector, $W \in \mathbb{R}^{N \times M}$ is the weight matrix, $b \in \mathbb{R}^N$ is the bias vector, and act is a pointwise activation function (for example the identity function or the ReLU function $\gamma \mapsto \max(0, \gamma)$). Entries in $W$ and $b$ are usually parameters to be learned in the training process.

In the training process, a network will go through iterations of forward and backward propagations. Suppose the loss function is denoted by $L$. In a forward propagation, one calculates $y$ based on a given $x$. In the backward propagation, one calculates $\partial L/\partial W$, $\partial L/\partial x$ and $\partial L/\partial b$ based on a given $\partial L/\partial y$. Suppose that $v$ is a learnable parameter for this layer. Then after one backward propagation with learning rate $r$, $v$ will be updated by

$$v - r\frac{\partial L}{\partial v} \mapsto v,$$

where $\partial L/\partial v$ will be calculated by using $\partial L/\partial W$, $\partial L/\partial x$ and $\partial L/\partial b$.

For any variable $z$ in a layer we use $z^{(0)}$ to denote its value after initialization, and we use $z^{(1)}$ to denote its value after the first forward and backward propagation. We introduce a definition here.

**Definition 2.1** Assuming the notations and single layer structure as in (1), let $z$ be a variable in this layer. Then the *increment of* $z$, denoted as $\Delta z$, is defined to be

$$\Delta z = z^{(1)} - z^{(0)}$$

when we set the learning rate to be 1.

For a learnable variable $v$ we have $\Delta v = -\partial L/\partial v$. For an intermediate variable $W_{nm}$ which corresponds to the learnable variable $v$ in the form $W_{nm} = cv$, if we assume that $c$ is a constant scalar (non-learnable) and $v$ only appears in $W$ in this layer, we have

$$\begin{aligned}\Delta W_{nm} &= W_{nm}^{(1)} - W_{nm}^{(0)} \\ &= cv^{(1)} - cv^{(0)} \\ &= -c\frac{\partial L}{\partial v} \\ &= -c^2 \sum_{(n',m')} \frac{\partial L}{\partial W_{n'm'}}, \end{aligned} \tag{2}$$

where the last summation is over all entries in $W$ that share the same learnable parameter $v$. If $v$ also appears in weights of other layers, the last summation need also to include the corresponding partial derivatives.

## 2.2 *CirCNN Networks*

Most state-of-the-art neural networks contain enormous amounts of learnable parameters. For example, the VGG network introduced in [23], which achieved the 1st place in 2014 ImageNet Challenge, contains more than one million learnable parameters. How to find efficient ways to reduce the number of parameters in deep neural networks is an active research area [8, 16]. A promising approach is to replace the matrices and convolutions in a deep neural network by structured matrices and structured convolutions. This approach has the advantage that the architecture of the network can be preserved. In [4] the authors proposed to replace unstructured matrices in a network by block-circulant matrices. In this paper we call any network which uses the technique in [4] as a CirCNN network. The idea is best illustrated via an example. We say a matrix is a *circulant matrix* if it has the form

$$\begin{bmatrix} a_1 & a_2 & \cdots & a_{n-1} & a_n \\ a_n & a_1 & \cdots & a_{n-2} & a_{n-1} \\ \vdots & \vdots & \vdots & \vdots & \vdots \\ a_2 & a_3 & \cdots & a_n & a_1 \end{bmatrix}.$$

Note that unlike an unstructured matrix, a circulant matrix is determined if one knows the first row (or first column) of the matrix. For a fully connected layer of the form $y = \text{act}(Wx + b)$, we say $W$ is a *block-circulant matrix* if $W$ consists of block matrices, where each block matrix is a circulant matrix. For a fully connected layer, suppose we replace an unstructured matrix by a block-circulant one with block size $B$. Then for this matrix we can reduce the amount of parameters by a factor of $B$. We will call this number $B$ the *compression ratio* of this layer. Note that for one layer, the number of parameters in the weight matrix $W$ will usually dominate the number of parameters in this layer. We also note that by using a block-circulant matrix, one can speed up matrix vector multiplication by exploiting the special structure of such a matrix [20].

Compressing a neural network by using block-circulant matrices can significantly reduce the amount of parameters and thus reduce bandwidth requirement. It is proved in [26] that the Universal Approximation Property is preserved if one replaces unstructured matrices by block-circulant ones. This guarantees the expressive power of CirCNN implementations.

## 3 The Xavier/Kaiming Initialization and its Limitation

In this section we briefly introduce the Xavier/Kaiming initialization. We give examples of networks to show that the Xavier/Kaiming initialization may not be sufficient to lead to fast and stable trainings.

For a fully connected layer of the form (1), entries in the matrix $W$ and the vector $b$ are the learnable parameters. Denote the final loss function as $L$, then the Xavier/Kaiming initialization has the following requirements

$$\begin{cases} \text{mean}(Y) = \text{mean}(X) = 0, \\ \text{var}(Y) = \text{var}(X), \\ \text{mean}(\frac{\partial L}{\partial Y}) = \text{mean}(\frac{\partial L}{\partial X}) = 0, \\ \text{var}(\frac{\partial L}{\partial Y}) = \text{var}(\frac{\partial L}{\partial X}), \end{cases} \tag{3}$$

where we assume that all entries in $x$ follow the distribution $X$, all entries in $y$ follow the distribution $Y$, all entries in $\frac{\partial L}{\partial x}$ follow the distribution $\frac{\partial L}{\partial X}$, and all entries in $\frac{\partial L}{\partial y}$ follow the distribution $\frac{\partial L}{\partial Y}$. Unless $W$ is a square matrix and the activation function is a linear function, the above conditions cannot be satisfied simultaneously, but the following initialization rules are widely used (see [11] for details): the bias vector $b$ should be initialized as the zero vector, and entries in $W$ should be identically and independently distributed with mean 0 and variance $\frac{2 \cdot \text{gain}}{M+N}$, where gain is a factor determined by the activation function. For the identity function, gain should be 1, while for the ReLU activation, gain should be 2. In practice the distribution for entries in $W$ is chosen as either the uniform distribution or the normal distribution. We note that the Xavier/Kaiming initialization conditions in (3) can maintain a constant variance for different layers in both the forward and the backward propagations.

We first consider a simple network with no bias of the form

$$y = W_2(\mathrm{ReLU}(W_1 x)), \tag{4}$$

where $x \in \mathbb{R}^M$ is the input vector, and $W_1$ and $W_2$ are the parameter matrices of sizes $\mathbb{R}^{M\times M}$ and $\mathbb{R}^{N\times M}$ respectively. Entries in the parameter matrices are learnable parameters. We can decompose the above network in two ways, each with two layers, as follows

$$\begin{cases} y_1 = \mathrm{ReLU}(W_1 x), \\ y_2 = W_2 y_1, \end{cases} \tag{5a}$$

$$\begin{cases} y_1 = \mathrm{ReLU}(cW_1 x), \\ y_2 = \frac{1}{c} W_2 y_1, \end{cases} \tag{5b}$$

where $c$ is a fixed positive scalar. With the same $W_1$, $W_2$ and $x$ these two decompositions will give the same final output $y_2$. By applying Xavier/Kaiming initializations to both networks, we can ensure that for both networks the mean and variance of the input and output of each layer are the same, and the mean and variance of the partial derivatives of the input and output of each layer are also the same. Routine calculations show that we should initialize as follows

$$\begin{cases} \mathrm{var}(W_1) = 2/M, \\ \mathrm{var}(W_2) = 2/(M+N), \end{cases} \tag{6a}$$

$$\begin{cases} \mathrm{var}(W_1) = 2/(c^2 M), \\ \mathrm{var}(W_2) = 2c^2/(M+N). \end{cases} \tag{6b}$$

We see that when $c = 1$, the two decompositions and initializations are the same.

To test the effect of the positive scalar $c$, we test the above two networks on the MNIST dataset. In this case $M = 784$ and $N = 10$. We set $c = 0.01$, and we use SGD (without moment) as the optimizer. The outputs $y_2$ of both networks are connected to softmax layers, and we use the cross entropy as the lost function in both cases (for details see [7]). Please refer to Fig. 1 for the experiment results. The learning rates for the two cases are tuned by trial and error to find the best value. We observed that the training of the second network (with $c = 0.01$) is much difficult than the training of the first network. As a result we need to use a much smaller learning rate, which results in slow convergence and a higher final loss value. To help visualize the difficulty in training the $c = 0.01$ case, in Fig. 1b, c we also plot distributions of $W_1$

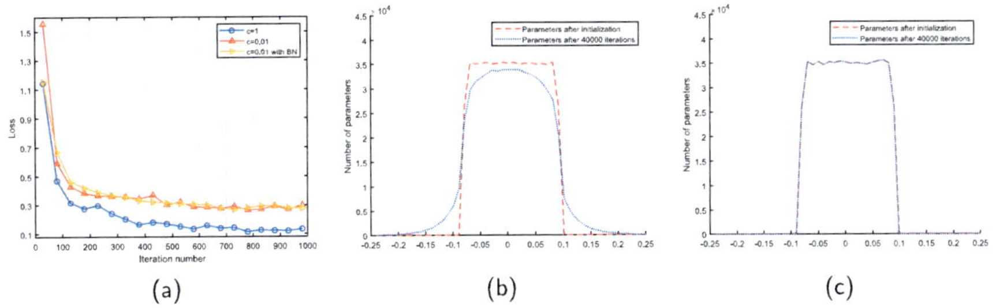

**Fig. 1** **a** Comparison of decompositions (5a) (represented by $c = 1$ in the graph) and (5b) with initializations (6a) and (6b). Plots start from the 31st iteration for better illustration of differences of curves. **b** Plot of distributions of $W_1$ in (5a) after initialization and 40000 iterations. **c** Plot of distributions of $cW_1$ in (5b) after initialization and 40000 iterations

in (5a) and distributions of $cW_1$ in (5b). It is clear that for the $c = 0.01$ case, there is no observable change in the distributions. This indicates that parameters in $W_1$ do not learn after iterations. This is because of the very small learning rate we have to use and the small value $c = 0.01$. On the other hand, for the $c = 1$ case, the distribution evolves from the uniform distribution to a bell-curved distribution. This indicates a successful learning process for parameters in $W_1$. We also noticed that with $c = 0.01$, the unstable training process cannot be remedied by batch normalization, which is introduced in [14] and is widely used to remedy unstable training.

We also tested the effect of the $c$ scalar and batch normalization (BN) on the VGG16 network with the Cifar10 dataset. We multiply $c$ on the first fully connected layer of VGG16, and then divide $c$ on the second fully connected layer. The test results on the validation set are summarized in Table 1. For all tests we used the Xavier/Kaiming initialization. For $c = 100$ with no BN, we have to reduce learning rate to get a good training, while for $c = 1000$ with no BN the training always fails (test accuracy can never reach 50%, which is too low for VGG16 on Cifar10). For these two $c$ values with BN layers, we notice that the VGG16 network can have stable training with a higher learning rate. However we still have about 1% accuracy loss. To learn why this happens, in Fig. 2 we plot standard deviations (std) of the parameters in weight matrices of the first fully connected layers of the original VGG16 and the $c = 1000$ case with BN. For the original VGG16, the std almost keeps constant. This is not surprising, as the Xavier/Kaiming initialization, when applied to a "normal" network, can stabilize statistics of learnable parameters. For VGG16 with $c = 1000$, we observe that the std first increases rapidly and then stabilizes. The stabilized std is much higher than the std of the parameters just after initialization by the Xavier/Kaiming method. For the $c = 1000$ case, the rapid change in std of learnable parameters and large deviation from initialized values are undesirable, since the initialized values satisfy the Xavier/Kaiming initialization which can maintain a constant variance for different layers in both the forward and the backward propagations, see [6, 11].

**Table 1** VGG16 network with Cifar10. The constant $c$ is multiplied to the first fully connected layer, and then divided from the second fully connected layer. For $c = 1000$ without BN, we fail to find a suitable learning rate to obtain an accuracy $\geq 50\%$

| | VGG16 with no BN accuracy on validation set (best lr) | VGG16 with BN accuracy on validation set (best lr) |
|---|---|---|
| Original | $92.24\%(2.5 * 10^{-2})$ | $92.24\%(2.5 * 10^{-2})$ |
| $c = 100$ | $83.53\%(5 * 10^{-4})$ | $91.03\%(5 * 10^{-3})$ |
| $c = 1000$ | <50% | $91.24\%(5 * 10^{-3})$ |

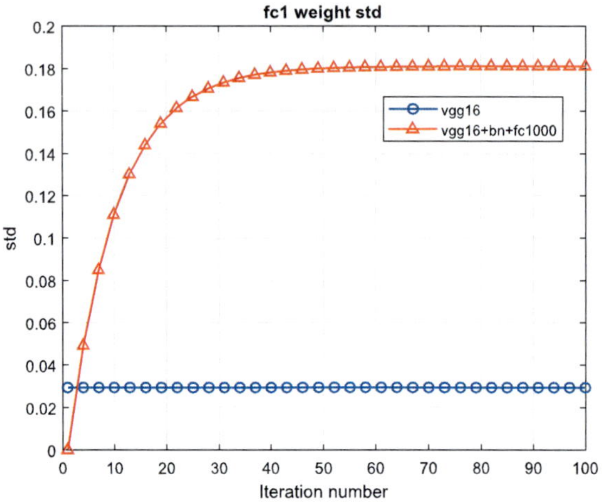


**Fig. 2** Plot of standard deviations of entries in the weight matrices of the first fully connected layers of the original VGG16 (blue line with circles) and the $c = 1000$ case (red line with triangles)

## 4 A New Initialization for Fully Connected Layers

In this section we describe our new initialization method for a fully connected layer. We will also give numerical experiments to show the effectiveness of this new method.

### *4.1 Our Initialization Conditions for Fully Connected Layers*

As demonstrated in the previous section, one must decompose a network properly in order to obtain stable training and good network performance. In this subsection we propose to add a new condition to the Xavier/Kaiming initialization for layers where the weight matrix does not have sparsity. This new condition can guarantee a more balanced initialization, and we will demonstrate how to decompose a network properly to satisfy the new condition.

We assume that a fully connected layer of a multi-layer network is of the form

$$\begin{cases} W_{nm} = \alpha v_{\lambda(n,m)}, \\ s_n = \sum_m W_{nm} x_m, \\ y_n = \mathrm{act}(s_n + b_n), \end{cases} \tag{7}$$

where $\alpha$ is a fixed positive scalar for this layer that will be determined later, $x \in \mathbb{R}^M$ is the input, $y \in \mathbb{R}^N$ is the output, $b \in \mathbb{R}^N$ is the bias, $v \in \mathbb{R}^\Lambda$ is the collection of learnable parameters, act is the activation function, and

$$\lambda : \{0, 1, \cdots, N\} \times \{0, 1, \cdots, M\} \to \{0, 1, \cdots, \Lambda\}$$

is a function that builds the matrix $W$ from the vector $v$. In this way we guarantee that each entry in $W$ corresponds to one element in $v$, but we allow multiple entries in $W$ to share the same entry in $v$. We always use $L$ to denote the loss function. We only consider SGD (without moment) as the optimizer. In our notation the learnable parameters are only contained in $v$ and $b$.

Assuming the notations and single layer structure in (7). We require that any initialization of (7) should satisfy the following conditions

$$\begin{cases} \mathrm{mean}(Y) = \mathrm{mean}(X) = 0, \\ \mathrm{var}(Y) = \mathrm{var}(X), \\ \mathrm{mean}(\frac{\partial L}{\partial Y}) = \mathrm{mean}(\frac{\partial L}{\partial X}) = 0, \\ \mathrm{var}(\frac{\partial L}{\partial Y}) = \mathrm{var}(\frac{\partial L}{\partial X}), \\ \mathrm{var}(\Delta W) = \mathrm{var}\left(\frac{\partial L}{\partial W}\right), \end{cases} \tag{8}$$

where $\partial L/\partial W$ represents the distribution followed by the entries $\partial L/\partial W_{nm}$.

We note that the first four conditions in (8) are exactly the Xavier/Kaiming initialization conditions (3). For a standard fully connected layer with $\alpha = 1$, the last condition in (8) will be satisfied automatically. The significance of the last condition will become apparent when we consider more general fully connected layers with parameter sharing, like the CirCNN fully connected layers introduced in Sect. 2.2.

For the splitting (5b) of (4) we have $\alpha = c$, and a simple calculation shows that (see (2))

$$\mathrm{var}(\Delta W) = \mathrm{var}\left(-c^2 \frac{\partial L}{\partial W}\right) = c^4 \cdot \mathrm{var}\left(\frac{\partial L}{\partial W}\right).$$

We see that in order to satisfy our initialization conditions (8), $c$ should be equal to 1. For $c = 0.01$, even though both initializations (6a) and (6b) satisfy the Xavier/Kaiming initialization, only the initialization (6a) satisfies our initialization conditions and it gives a better network decomposition.

## 4.2 New Initialization for CirCNN Fully Connected Layers

For the CirCNN network in [4], the fully connected layers can be expressed as (7) with $\alpha = 1$. A routine calculation shows that

$$\Delta W_{nm} = \Delta v_{\lambda(n,m)} = - \sum_{\lambda(n',m')=\lambda(n,m)} \frac{\partial L}{\partial W_{n'm'}}, \tag{9}$$

where the last summation is over all entries in $W$ that share the same learnable parameter with $W_{nm}$. We note that the weight sharing comes from the circulant matrices in $W$. We denote $V_{nm} = \{W_{n'm'} : \lambda(n', m') = \lambda(n, m)\}$, and we use $B_{nm}$ to denote the number of elements in $V_{nm}$. Assuming that

$$\left\{ \frac{\partial L}{\partial W_{n'm'}} : W_{n'm'} \in V_{nm} \right\}$$

is uncorrelated, from (9) we have

$$\begin{aligned} \mathrm{var}(\Delta W_{nm}) &= \sum_{W_{n'm'} \in V_{nm}} \mathrm{var}\left( \frac{\partial L}{\partial W_{n'm'}} \right) \\ &= B_{nm} \mathrm{var}\left( \frac{\partial L}{\partial W} \right). \end{aligned} \tag{10}$$

Then we see that the only way to satisfy the last condition in our initialization (8) is to have $B_{nm} = 1$. However, this corresponds to circulant matrices of size $1 \times 1$, which simply implies that the CirCNN does not compress the original network at all.

In order to achieve a desirable compression rate, we need to reformulate the fully connected layer with CirCNN in [4] in the form of (7) and determine a proper value for the extra positive scaler $\alpha$. We stress that the scalar $\alpha$ will be a constant for this layer, and the learnable variables are still only in $v$ and $b$. Thus by introducing $\alpha$, we do not increase the number of learnable parameters, and the increment in the number of operations for this layer is negligible.

Assume that we wish to compress this layer by a factor of $B$. Then the circulant sub-matrices in $W$ should have size $B \times B$. A calculation combining (2) and (10) shows that

$$\mathrm{var}(\Delta W_{nm}) = \mathrm{var}\left( \alpha^2 \sum \frac{\partial L}{\partial W_{n'm'}} \right) = \alpha^4 B_{nm} \mathrm{var}\left( \frac{\partial L}{\partial W} \right), \tag{11}$$

where $B_{nm} = B$. Then to satisfy the last initialization condition in (8), we can simply set $\alpha = 1/\sqrt[4]{B}$. Combining the Xavier/Kaiming initialization conditions, which constitute the first 4 conditions in (8), we conclude that our initialization leads to the following rule

$$\begin{cases} \mathrm{var}(v_{\lambda(n,m)}) = \mathrm{gain} \cdot \sqrt{B} \cdot \frac{2}{M+N}, \\ \alpha = \frac{1}{\sqrt[4]{B}}, \end{cases} \tag{12}$$

where gain should be determined by the activation function.

To sum it up, our new initialization method for CirCNN implemented fully connected layer consists of 3 steps. The first step is to write the layer in the form (7). The second step is to calculate $\alpha$ and variance by (12). The last step is to initialize entries in $v$ by independently and identically distributed variables with mean 0 and the variance calculated in step 2 (one can use either the uniform or the normal distribution), and initialize the bias vector $b$ as the zero vector.

To see the effect of the positive scalar $\alpha$ on the updates of $v_{\lambda(n,m)}$ we calculate as follows. Suppose the learning rate of the SGD optimizer is $r$, then after one forward and backward propagation $v_{\lambda(n,m)}$ is updated by

$$v_{\lambda(n,m)} - \alpha r \sum_{W_{n'm'} \in V_{nm}} \frac{\partial L}{\partial W_{n'm'}} \mapsto v_{\lambda(n,m)}. \tag{13}$$

We see that now the effective learning rate for $v_{\lambda(n,m)}$ changes from $r$ to $\alpha r$. Thus by formulating each layer as in (7), we can change the effective learning rate of learnable parameters layer by layer.

## 4.3 Numerical Experiments

We demonstrate the effectiveness of our initialization method by numerical experiments.

**Example 4.1** Here we show the training process of a simple network with the initialization (12) on the MNIST dataset. The network structure is summarized in Table 2, and we use CirCNN compression for the first fully connected layer. Note that after the compression, the total number of learnable parameters is only about 0.76% of the original model. From the plot of the loss function in Fig. 3a, we see that our initialization method better suits this CirCNN network.

**Example 4.2** In this example we consider a simple network where the weight matrix of many layers share a common weight matrix $V$. Details of the network, together with the initialization of the common weight $V$ by our method (8), is summarized in (14a) and (14b)

**Table 2** Network structure of Example 4.1. For a compression ratio $B \geq 1$, we use circulant implementation with block size $B$. The number of parameters for a CirCNN implementation should be divided by $B$

| Layer | Output channel | kernel/matrix size | Compression ratio |
|---|---|---|---|
| Conv2d | 32 | $3 \times 3 \times 1 \times 32$ | 1 |
| MaxPool | | | |
| Conv2d | 64 | $3 \times 3 \times 32 \times 64$ | 1 |
| MaxPool | | | |
| Fc | 1568 | $3136 \times 1568$ | 1568 |
| Fc | 10 | $1568 \times 10$ | 1 |

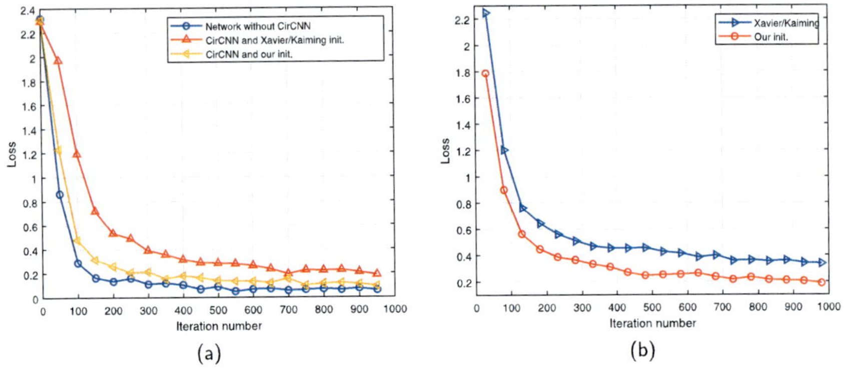


**Fig. 3** **a** Plot of losses of networks summarized in Table 2. **b** Plot of losses of network summarized in (14a). For (**b**), the plot starts from the 31st iteration for better illustration of differences of curves

$$\begin{cases} y_0 = \mathrm{ReLU}(W_0 x), \\ W_1 = W_2 = \cdots = W_{50}, \\ y_1 = W_1 y_0, \\ \vdots \\ y_{50} = W_{50} y_{49}, \\ y = W y_{50}, \end{cases} \tag{14a}$$

$$\begin{cases} \alpha = 1/\sqrt{50}, \\ \mathrm{var}(V) = 1/(\alpha^2 M), \end{cases} \tag{14b}$$

where as before, for our initialization the weight matrices have the form $W_i = \alpha V$ for $i = 1, 2, \cdots, 50$. A comparison with our initialization (14b) and the Xavier/Kaiming initialization is summarized in Fig. 3b.

**Table 3** CirCNN implementation of VGG16 on Cifar10. Our initialization method consistantly outperforms the Xavier/Kaiming initialization. Reported results are top-1 accuracy. Original VGG16 can achieve 92.24% top-1 accuracy

| | Number of runs | | | | | | |
|---|---|---|---|---|---|---|---|
| | lr | 1 | 2 | 3 | 4 | 5 | Mean |
| Xavier/Kaiming | $10^{-2}$ | 91.19% | 91.08% | 90.83% | 90.81% | Fail to converge | 90.98% |
| | $5 * 10^{-3}$ | 89.61% | 89.69% | 89.41% | 86.19% | 90.10% | 89.00% |
| Our method | $2.5 * 10^{-2}$ | 92.36% | 92.02% | 91.99% | 91.85% | 91.80% | 92.00% |

**Example 4.3** Lastly we test our initialization method on a CirCNN implementation of the VGG16 network. For the second fully connected layer in the VGG16 network which has 4096 input channels and 4096 output channels, we use a circulant matrix with block size 4096. We test on the Cifar10 dataset with 5 runs for both Xavier/Kaiming and our initialization methods. The original VGG16 network can achieve 92.24% top-1 accuracy. The test results are summarized in Table 3. On the 5 runs, the network with our initialization consistently outperforms the network with the Xavier/Kaiming initialization on top-1 accuracy. Because of the CirCNN implementation, we have to lower learning rates for the Xavier/Kaiming initialization. With a learning rate of $10^{-2}$, one training with the Xavier/Kaiming initialization failed to converge. On the other hand, the learning rate for our initialization is $2.5 * 10^{-2}$ which is identical to the learning rate of the original VGG16 network.

## 5 Conclusion and Future Work

In this paper we identified a limitation of the widely used Xavier/Kaiming initialization method, and we proposed an efficient initialization method for fully connected layers. We also proposed a method to adjust the learning rate of parameters in each layer. Our new initialization method is easy to implement. For networks with heavy weight sharing, experiments show that our new method has a clear advantage over the Xavier/Kaiming initialization method.

## References

1. O. Abdel-Hamid, A.-R. Mohamed, H. Jiang, L. Deng, G. Penn, D. Yu, Convolutional neural networks for speech recognition. IEEE/ACM Trans. Audio Speech Lang. Process. **22**(10), 1533–1545 (2014)
2. R. Collobert, J. Weston, L. Bottou, M. Karlen, K. Kavukcuoglu, P. Kuksa, Natural lan- guage processing (almost) from scratch. J. Mach. Learn. Res., 2493–2537 (2011)

3. J. Devlin, M.-W. Chang, K. Lee, K. Toutanova, Bert: pre-training of deep bidirectional transformers for language understanding. In: arXiv preprint arXiv:1810.04805 (2018)
4. C. Ding, S. Liao, Y. Wang, Z. Li, N. Liu, Y. Zhuo, C. Wang, X. Qian, Y. Bai, G. Yuan, et al., Cir- CNN: accelerating and compressing deep neural networks using block-circulant weight matrices, in *Proceedings of the 50th Annual IEEE/ACM International Symposium on Microarchitecture*. ACM, pp. 395–408 (2017)
5. B. Fulkerson, A. Vedaldi, S. Soatto, Class segmentation and object localization with superpixel neighborhoods, in *IEEE 12th International Conference on Computer Vision*. IEEE, pp. 670–677 (2009)
6. X. Glorot, Y. Bengio, Understanding the difficulty of training deep feedforward neural networks, in *Proceedings of the Thirteenth International Conference on Artificial Intelligence and Statistics* pp. 249–256 (2010)
7. I. Goodfellow, Y. Bengio, A. Courville, *Deep Learning*. http://www.deeplearningbook.org. MIT Press (2016)
8. S. Han, X. Liu, H. Mao, J. Pu, A. Pedram, M.A. Horowitz, W.J. Dally, EIE: efficient inference engine on compressed deep neural network, in *ACM/IEEE 43rd Annual Inter- national Symposium on Computer Architecture (ISCA)*, pp. 243–254 (2016)
9. B. Hanin, D. Rolnick, How to start training: The effect of initialization and architecture, in *Advances in Neural Information Processing Systems*, pp. 571–581 (2018)
10. K. He, X. Zhang, S. Ren, J. Sun, Deep residual learning for image recognition, in *The IEEE Conference on Computer Vision and Pattern Recognition (CVPR)* (2016)
11. K. He, X. Zhang, S. Ren, J. Sun, Delving deep into rectifiers: surpassing human-level performance on ImageNet classiffication, in *The IEEE International Conference on Computer Vision (ICCV)* (2015)
12. G. Hinton, L. Deng, D. Yu, G. Dahl, A.-R. Mohamed, N. Jaitly, A. Senior, V. Vanhoucke, P. Nguyen, T. Sainath, B. Kingsbury, Deep neural networks for acoustic modeling in speech recognition. IEEE Signal Process. Mag., 29 (2012)
13. S. Hochreiter, J. Schmidhuber, Long short-term memory. Neural Comput. **9**(8), 1735–1780 (1997)
14. S. Ioffe, C. Szegedy, Batch normalization: accelerating deep network training by reducing internal covariate shift, in *International Conference on Machine Learning*, pp. 448–456 (2015)
15. A. Krizhevsky, I. Sutskever, G.E. Hinton, Imagenet classiffication with deep convolutional neural networks, in *Advances in Neural Information Processing Systems*, pp. 1097–1105 (2012)
16. J.-H. Luo, J. Wu, W. Lin, Thinet: a filter level pruning method for deep neural network compression, in *Proceedings of the IEEE International Conference on Computer Vision*, pp. 5058–5066 (2017)
17. T. Mikolov, M. Karafiát, L. Burget, J.Černocỳ, S. Khudanpur, Recurrent neural network based language model, in *Eleventh Annual Conference of the International Speech Communication Association* (2010)
18. D. Mishkin, J. Matas, All you need is a good init, in *International Conference on Learning Representations* (2016)
19. M. Mostajabi, P. Yadollahpour, G. Shakhnarovich, Feedforward semantic segmentation with zoom-out features, in *Proceedings of the IEEE Conference on Computer Vision and Pattern Recognition*, pp. 3376–3385 (2015)
20. V. Y. Pan, *Structured Matrices and Polynomials: Unified Superfast Algorithms* Springer (Science & Business Media, 2012)
21. A. Saxe, J. L. McClelland, S. Ganguli, Exact solutions to the nonlinear dynamics of learning in deep linear neural networks. In: arXiv preprint arXiv:1312.6120 (2013)
22. D. Silver, A. Huang, C.J. Maddison, A. Guez, L. Sifre, G. Van Den Driessche, J. Schrittwieser, I. Antonoglou, V. Panneershelvam, M. Lanctot, et al., Mastering the game of Go with deep neural networks and tree search. Nature **529**(7587), 484 (2016)
23. K. Simonyan, A. Zisserman. "Very deep convolutional networks for large-scale image recognition. In: arXiv preprint arXiv:1409.1556 (2014)

24. L. Xiao, Y. Bahri, J. Sohl-Dickstein, S. Schoenholz, J. Pennington, Dynamical isometry and a mean field theory of CNNs: How to train 10,000-layer vanilla convolutional neural networks, in *International Conference on Machine Learning*, pp. 5389–5398 (2018)
25. G. Yang, S. Schoenholz, Mean field residual networks: on the edge of chaos, in *Advances in neural Information Processing Systems*, pp. 7103–7114 (2017)
26. L. Zhao, S. Liao, Y. Wang, Z. Li, J. Tang, B. Yuan, Theoretical properties for neural networks with weight matrices of low displacement rank, in *Proceedings of the 34th International Conference on Machine Learning* Vol. 70. JMLR. org, pp. 4082–4090 (2017)

# The Shortest Path AMID 3-D Polyhedral Obstacles

**Shui-Nee Chow, Jun Lu, and Hao-Min Zhou**

**Abstract** It is well known that the problem of finding the shortest path amid 3-D polyhedral obstacles is a NP-Hard problem. In this paper, we propose an efficient algorithm to find the globally shortest path by solving stochastic differential equations (SDEs). The main idea is based on the simple structure of the shortest path, namely it consists of straight line segments connected by junctions on the edges of the polyhedral obstacles. Thus, finding the shortest path is equivalent to determining the junctions points. This reduces the originally infinite dimensional problem to a finite dimensional one. We use the gradient descent method in conjunction with Intermittent Diffusion (ID), a global optimization strategy, to deduce SDEs for the globally optimal solution. Compared to the existing methods, our algorithm is efficient, easier to implement, and able to obtain the solution with any desirable precisions.

**Keywords** Path planning · Stochastic differential equations · Intermittent diffusion · Obstacle avoidance · Global optimization

---

Dedicated to Professor RAYMOND H. CHAN on the occasion of his 60th birthday

---

This work is partially supported by NSF Awards DMS-1419027, DMS-1620345, and ONR Award N000141310408

---

S.-N. Chow · J. Lu · H.-M. Zhou (✉)
School of Mathematics, Georgia Institute of Technology, Atlanta, GA 30332, USA
e-mail: hmzhou@math.gatech.edu

S.-N. Chow
e-mail: chow@math.gatech.edu

J. Lu
e-mail: junlu0@icloud.com



X.-C. Tai et al. (eds.), *Mathematical Methods in Image Processing and Inverse Problems*, Springer Proceedings in Mathematics & Statistics 360,
https://doi.org/10.1007/978-981-16-2701-9_10

# 1 Introduction

Finding the shortest path in the presence of obstacles is one of the fundamental problems in path planning and robotics. It is one of the enabling technologies that make it possible for robots or UAVs to traverse cluttered environments. The problem can be described as follows: given a finite number of obstacles in $\mathbf{R}^2$ or $\mathbf{R}^3$, what is the shortest path connecting two given points $X, Y$ while avoiding the obstacles. The problem has received great attention during the last few decades (See for example [6, 12] and references therein), and many techniques have been developed for polygonal obstacles in $\mathbf{R}^2$, where the problem can be reformulated as an optimization problem on a graph, and therefore can be solved by combinatorial methods. For example, by using the shortest path map method, Hershberger and Suri [9] found an optimal $\mathbf{O}(n \log n)$ polynomial time algorithm where $n$ is the total number of vertices of all polygonal obstacles. We refer to [12, 14] for a survey of the results and references therein. However, Canny and Rief [2] proved that this problem in $\mathbf{R}^3$ becomes NP-hard under the framework known as "configuration space". This is mainly because the shortest path doesn't necessarily pass through the set of vertices of polyhedrons. Instead, it may go through the interior points of edges, and this makes the optimal algorithm in 2-D fail.

Two different approaches were developed later to overcome this difficulty. One is to find a path that is $1 + \epsilon$ times the length of the shortest one. The idea is to subdivide the edges in certain ways and adopt the same optimal combinatorial methods which are effective in $\mathbf{R}^2$. Following this idea, Papadimitriou developed an algorithm in [15] with complexity $O(\frac{1}{\epsilon})$. In a special case where the shortest path is unique, one can define the precision $\delta$ of the problem, which is the difference between the shortest path and the second shortest path. Given $\epsilon < \delta$, a faster algorithm was developed in [3] with complexity $O(\log(\frac{1}{\epsilon}) + P(1/\delta))$ for some polynomial $P$. The idea is to apply the approach in [15] to obtain a good initialization within error $\delta$ to the shortest path, and then use a gradient descent strategy to improve the accuracy to $\epsilon$.

Another commonly used approach divides the problem into two parts: (i) find the sequence of edges that the shortest path may go through, and (ii) find the optimal connecting points on those edges. For convex polyhedral obstacles, it is observed in [19] that the total number of possible sequences are of order $O(n^{7k}k^k)$ where $n$ is the total number of vertices and $k$ is the number of obstacles. Part (ii) is proven to be NP-hard [7]. A different method, called unfolding technique, was introduced in [17] under a theoretical computation model in which it assumes any infinite-precision real arithmetic operation requires constant time. However, this assumption may not be practical.

On the other hand, several differential equation based methods have been proposed to tackle the shortest path problem with obstacles having smooth boundaries. For example, a path evolution method finds the solution by solving a 2-point boundary value ordinary differential equation (ODE), resulting in locally optimal solutions. The front propagation method finds the global solution by solving an eikonal equation, a partial differential equation (PDE). The numerical solution of the eikonal equation can be computed by the fast marching method [18] or the fast sweeping method [20].

In [5], we proposed a different algorithm called Evolving Junctions on Obstacle Boundaries (E-JOB) for finding the shortest path. E-JOB is a general framework which can be applied to environment with obstacles of arbitrary shape (continuous or discrete) in any dimension ($\mathbf{R}^2$, $\mathbf{R}^3$ or higher). The key idea is dimension reduction. It takes advantage of a simple geometric structure of the shortest path, i.e. the shortest path is composed by line segments and arcs on the obstacle boundaries. The shortest path is determined completely by the junctions of those segments. In this way, the problem becomes how to find those junction points on the boundaries. In other words, the original infinite dimensional problem of finding the whole path is converted to a finite dimensional problem of finding only the junction points. The optimal position of those junctions can be determined efficiently by the gradient descent method. To address the drawback that the gradient descent method usually gets stuck at local minimizers, a global optimization strategy called intermittent diffusion (ID) is adopted. This strategy adds random perturbations to the ODEs of the gradient descent method in a temporally discontinuous fashion, which leads to stochastic differential equations (SDEs). It obtains the globally shortest path with probability $1 - \delta$ where $\delta$ is an arbitrarily small number. More specifically, by leveraging the recent studies of convergence rate of Fokker-Planck equations [1, 4, 10, 13], it has been shown that the time complexity of E-JOB is $O(\log \frac{1}{\delta} \log \frac{1}{\epsilon})$.

In this paper, we focus on applying E-JOB to the shortest path problem with polyhedral obstacles in $\mathbf{R}^3$. The restriction on polyhedral obstacles allows us to achieve further dimension reductions. For obstacles with smooth boundaries, the implementation of E-JOB requires computations of geodesic on the boundaries between two given points. In [5], this is achieved by either traversing the boundaries in $\mathbf{R}^2$, or fast marching on the boundary surfaces in $\mathbf{R}^3$. However, for polyhedral obstacles, the geodesics also has a similar simple structure, i.e. the geodesic between two points on a polyhedron is a concatenation of line segments whose ending points are located on polyhedron edges. And to determine the geodesic is equivalent to determining those junction points. Therefore the overall shortest path connecting $X$ and $Y$ is merely a conjunction of line segments whose ending points lie on obstacle edges. In other words, each junction moves in a 1-D interval(edge). This makes the algorithm extremely simple and efficient.

A feature of this study is that we do not restrict the obstacles to be convex polyhedrons. The algorithm we develop can equally be applied to non-convex polyhedrons. For polyhedrons with Euler characteristic 2, which include all convex polyhedrons and concave polyhedrons without holes, our algorithm can find the globally optimal path with probability arbitrarily close to 1 in finite time. However, when dealing with more sophisticated polyhedrons, for example, polyhedrons with complicated holes, certain topological problems emerge, and prevent us from obtaining the globally optimal path. We will discuss this issue at the end of the paper as well as some possible solutions.

It should be noted that our approach resembles the one in [3] in the sense that both employ a gradient descent strategy. However, before the strategy can be applied, the method in [3] requires the assumption that the shortest path is unique, and an initialization that approximates the shortest path within error $\delta$ (precision) to start with (achieved by using [15]). Our approach needs neither of them. In fact, our approach can be viewed as a way to find both the edge sequence and the optimal connecting points in a unified manner, thanks to the introduction of randomness into the differential equations. Below, we summarize some advantages of our algorithm:

(1) The algorithm can obtain the shortest path in any precision. This is because only a system of SDEs needs to be solved which involves no subdivision of edges.
(2) The algorithm is able to handle non-convex polyhedral obstacles.
(3) The algorithm is easy to implement.
(4) The algorithm is fast. Since we solve an initial value problem of SDEs, the results can be obtained efficiently by various established schemes.

The paper is arranged as follows. In Sect. 2, we give the derivation of the algorithm following the ideas presented in [5]. The algorithm is then presented whose details follow afterwards. In Sect. 3, we give several interesting examples. Finally, we discuss the topological issues when dealing with polyhedrons with holes.

## 2 New Algorithm

In this section, we present our new algorithm for the shortest path problem with polyhedron obstacles. We start with some mathematical description of the problem, through which we introduce notations needed in the rest of the paper. The algorithm follows afterwards and its details are presented at the end of this section.

Let $\{P_k\}_{k=1}^N$ be $N$ polyhedral obstacles in $\mathbf{R}^3$. Each obstacle $P_k$ is determined uniquely by its vertices, edges and faces. Denote $V$, $E$, $F$ the set of vertices, edges and faces of $P_k$ respectively. We do not limit the polyhedrons to be convex. However, we will focus on polyhedrons without holes in this section, i.e. polyhedrons whose Euler characteristic is 2. The Euler characteristic is defined by

$$\chi = |V| - |E| + |F|.$$

Polyhedrons with holes will be discussed in the last section. For any edge $e \in E$, it has a representation

$$e = (\mathbf{u}, \mathbf{v})$$

where $\mathbf{u}, \mathbf{v}$ are the coordinates of the ending points of $e$. Any point $x$ on edge $e = (\mathbf{u}, \mathbf{v})$ can then be represented by the following expression

$$x(\mathbf{u}, \mathbf{v}, \theta) = \theta \mathbf{u} + (1 - \theta)\mathbf{v}, \tag{1}$$

where $\theta$ is a scalar in $[0, 1]$. Thus to determine the position of a point on an edge, one only needs to find its corresponding $\theta$.

## 2.1 Geodesics on Polyhedrons

For any two points $x, y$ on the edges of $P_k$, we can define the distance $d_k(x, y)$ between them to be the length of the shortest path on $P_k$ connecting $x$ and $y$. If we view $P_k$ as a surface in $\mathbf{R}^3$, i.e. a two dimensional manifold, then $d_k(x, y)$ is nothing but length of the geodesic on $P_k$ connecting $x$ and $y$. For instance, for any $x$ and $y$ on the same surface of a tetrahedron, $d(x, y) = \|x - y\|$ since the line segment joining them is on the surface. For general polyhedrons, the shortest path is composed by a sequence of line segments connected to each other. To be more specific, the shortest path can be represented by $(x_0, x_1, x_2, \ldots, x_{n_k}, x_{n_k+1})$ where $x_0 = x, x_{n_k+1} = y$ and each $x_i$ is a point on some edge $e_i = (\mathbf{u}_i, \mathbf{v}_i)$. The shortest distance $d_k(x, y)$ therefore equals

$$d_k(x, y) = \mathcal{L}(x_1, x_2, \ldots, x_{n_k}) = \sum_{i=0}^{n_k} \|x_{i+1} - x_i\|.$$

Denote $x_i = \theta_i \mathbf{u}_i + (1 - \theta_i)\mathbf{v}_i$, we then have

$$\mathcal{L}(\theta_1, \ldots, \theta_{n_k}) = \mathcal{L}(x_1, \ldots, x_{n_k}) = \sum_{i=0}^{n_k} \|\theta_{i+1}\mathbf{u}_{i+1} + (1 - \theta_{i+1})\mathbf{v}_{i+1} - \theta_i \mathbf{u}_i - (1 - \theta_i)\mathbf{v}_i\|.$$

It is worth mentioning that both $\theta$ and $\mathbf{u}_i, \mathbf{v}_i$ are dynamic as we optimize over $x_i$s.

## 2.2 Structure of the Shortest Path

A path is a curve $\gamma \in \mathbf{R}^3$, which is a continuous map

$$\gamma(\cdot)\colon [0, 1] \to \mathbf{R}^3.$$

We denote $L(\gamma)$ the Euclidean length of the path $\gamma$. We are concerned with the set of feasible paths $\mathbf{F}$, i.e. paths that do not intersect with any obstacle $P_k$. The shortest path connecting $X$ and $Y$ is then given by

$$\gamma_{opt} = \text{argmin}_{\gamma \in \mathbf{F}} L(\gamma).$$

In [5], we proved that the shortest path has a simple structure, i.e. it is composed by line segments outside the obstacles and paths on the boundary of the obstacles. Since all the obstacles here are polyhedrons, the paths on the boundaries of the obstacles also consist of a sequence of line segments connected by points on the edges. Therefore, by putting all the connecting points together and relabeling them, the shortest path connecting $X$ and $Y$ can be represented by $(x_0, x_1, x_2, \ldots, x_n, x_{n+1})$ where $x_0 = X, x_{n+1} = Y$.

Let us denote

$$J(x_i) = \|x_{i-1} - x_i\| + \|x_{i+1} - x_i\|. \tag{2}$$

Then the length of the path is

$$\mathcal{L}(x_1, \ldots, x_n) = \frac{1}{2}\sum_{i=1}^{n}(J(x_i) + \|x_1 - x_0\| + \|x_{n+1} - x_n\|). \tag{3}$$

Again all $x_i$s are on the edges of the obstacles. Denote $x_i = \theta_i \mathbf{u}_i + (1 - \theta_i)\mathbf{v}_i$, $J(x_i)$ then becomes

$$J(\theta_i) = \|\theta_i \mathbf{u}_i + (1 - \theta_i)\mathbf{v}_i - x_{i-1}\| + \|\theta_i \mathbf{u}_i + (1 - \theta_i)\mathbf{v}_i - x_{i+1}\|. \tag{4}$$

## 2.3 Optimal Path

To find the optimal path, we differentiate $J(\theta_i)$ with respect to $\theta_i$ to obtain

$$\nabla J(\theta_i) = \frac{(x_i - x_{i-1}) \cdot (\mathbf{u}_i - \mathbf{v}_i)}{\|x_i - x_{i-1}\|} + \frac{(x_i - x_{i+1}) \cdot (\mathbf{u}_i - \mathbf{v}_i)}{\|x_i - x_{i+1}\|}. \tag{5}$$

So using the method of gradient decent, we can find the optimal position $\theta_i$ following a system of ODEs,

$$\frac{d\theta_i}{dt} = -\nabla J(\theta_i). \tag{6}$$

In order to find the globally optimal path, we adopt a strategy called Intermittent Diffusion, i.e. we evolve the following SDE

$$\frac{d\theta_i}{dt} = -\nabla J(\theta_i) + \sigma(t)dW(t) \tag{7}$$

where $\sigma(t)$ is a step function and $W(t)$ is standard Brownian motion. More precisely,

$$\sigma(t) = \sum_{l=1}^{m} \sigma_l \mathbf{1}_{[S_l, T_l]}(t) \tag{8}$$

with $0 = S_1 < T_1 < S_2 < T_2 < \cdots < S_m < T_m < S_{m+1} = T$, and $\mathbf{1}_{[S_l, T_l]}$ being the indicator function of interval $[S_l, T_l]$. We note that adding $dW(t)$, the so-called white noise, to the SDE corresponds to having a diffusion process in the classical stochastic theory. Thus adding white noise perturbations to the SDE on discontinuous intervals $[S_l, T_l]$ is like adding diffusion process intermittently. For convenience, we call $[S_l, T_l$ an intermittent diffusion interval. In this paper, the intervals $S_l$, $T_l$, $\sigma_l$ are chosen to be random and the magnitude, $\{\sigma_l\}$, of the random perturbations are selected according to the following theorem. For more details, see [4].

**Theorem 1** *If all the obstacles have Euler characteristic 2, then for any small number $\epsilon$, there exists $\tau > 0$, $\sigma_0 > 0$ and integer $m_0 > 0$ such that if $T_i - S_i > \tau$, $\sigma_i < \sigma_0$ and $m > m_0$, then Eq. (7) converges to a global minimizer with probability $1 - \epsilon$.*

We will postpone the proof until the last section when we discuss the topological issues.

## 2.4 Numerical Scheme

In this section, we discuss how to solve Eq. (7).

### 2.4.1 Discretization of SDE

We use the forward Euler method to discretize equation (7)

$$\frac{\theta_i^{j+1} - \theta_i^j}{\Delta t} = -\nabla J(\theta_i^j) + \sigma(j\Delta t)\sqrt{\Delta t}\xi$$

where $\xi \sim N(0, 1)$ is a normal random variable. Notice the $\theta_i$s are updated alternatingly in the Gauss-Seidel fashion.

### 2.4.2 Initialization

We can use the optimal path whose junctions are restricted to vertices of the obstacles to initialize the path. This initialization can be obtained efficiently by a method called visibility graph . The visibility graph $W$ is a weighted graph whose nodes are the vertices of all the obstacles as well as the starting and ending points $X$, $Y$, and there is an edge between vertices $\mathbf{u} \in W$ and $\mathbf{v} \in W$ if and only if they are visible to each other, that is, if the line segment $\overline{\mathbf{uv}}$ doesn't intersect with any obstacles. The weight of edge $\mathbf{uv}$ is simply the Euclidean distance of $\overline{\mathbf{uv}}$. One thing to notice is that the visibility graph we construct here is essentially 2D, in the sense that it encodes whether two points are visible to each other. This is fundamentally different from the 3D reduced visibility graph (3DRVG) [11]. 3DRVG consists of connected planes as opposed to straight line segments which becomes complicated when there are more than one obstacles. After the visibility graph is constructed, the initialization is the shortest path between $X$ and $Y$ on the visibility graph $W$ which can be obtained efficiently by Dijkstra's algorithm.

### 2.4.3 Evolution when a Junction Reaches a Vertex

In the proposed method, the junctions move according to the SDEs if they are on the interior of edges. When a junction $x = (\mathbf{u}, \mathbf{v}, \theta)$ reaches a vertex $\mathbf{u}$ following the gradient flow, it continues moving according to different rules depending on whether the two neighbors of $x$ are on the same obstacle or not. If the neighbors of $x$ are both on the same obstacle as $x$, we call $x$ an interior junction, otherwise we call $x$ an exterior junction. In other words, an exterior junction is one of the two ending points of the line segments that connect two different obstacles. The following are the rules for interior and exterior junctions reaching the vertices respectively.

Case 1. $x = (\mathbf{u}, \mathbf{v}, \theta = 1)$ is an interior junction. As an example, see the following illustration (Fig. 1) where $(x_1, x_2, x, x_4)$ is the path on the obstacle and $x_1, x_4$ are exterior junctions. When $x$ hits $\mathbf{u}$ $(\theta = 1)$, path $(x_1, u = x, x_4)$ will have smaller length than $(x_1, x_2, u = x, x_4)$. In other words, all the junctions adjacent to $\mathbf{u}$ will be dragged to $\mathbf{u}$ except the exterior junctions. Hence we remove all the junctions adjacent to $\mathbf{u}$ and add junctions on the edges adjacent to $\mathbf{u}$ that haven't been occupied which results in a new path $(x_1, x_6, x_5, x_4)$.

Case 2. $x = (\mathbf{u}, \mathbf{v}, \theta = 1)$ is an exterior junction. Let $z$ be its neighbor on the same obstacle and $y$ be the neighbor on another obstacle. $x$ will move to a different feasible edge $\mathbf{uw}$ once it hits $\mathbf{u}$. Edge $\mathbf{uw}$ is said to be feasible if

(a) The line segment joining $y$ and $\mathbf{u} + \Delta\theta(\mathbf{w} - \mathbf{u})$ doesn't intersect with any obstacle for arbitrarily small $\Delta\theta$.

We collect all the feasible directions and select one of them with equal possibility, $x$ then continues evolving according to the flow. Depending on whether neighbor $z$ is visible, i.e. on the same face as edge $\mathbf{uw}$, the new path are as follows:

i. $z$ is on the same face as $\mathbf{uw}$, then the new path becomes $(\cdots, y, x', z, \cdots)$ where $x' \in \mathbf{uw}$.

ii. $z$ is not on the same face as $\mathbf{uw}$, then $x$ is used as an intermittent junction and the new path becomes $(\cdots, y, x', x, z, \cdots)$ where $x' \in \mathbf{uw}$.

For an illustration, see the following example (Fig. 2). The feasible directions are $\mathbf{uv}_8$ and $\mathbf{uv}_2$. $z$ is visible to $\mathbf{uv}_8$, the path after evolution is simply $(y, x', z)$. On the other hand, $z$ is invisible to $\mathbf{uv}_2$, the path after evolution is simply $(y, x', x, z)$.

Case 3. $x = (\mathbf{u}, \mathbf{v}, \theta = 1)$ is an exterior junction, and two of its neighbors, $z$ and $y$, are on other obstacles. There are two scenarios. One is when $x$ approaching $\mathbf{u}$, we remove $x$ from the junction list and add new points on the edges adjacent to $\mathbf{u}$ except $\mathbf{uv}$. This is the same scenario as that illustrated in Case 1. The other is that one can directly connect $z$ and $y$, and this involves adding and removing junctions as be discussed in the following subsection.

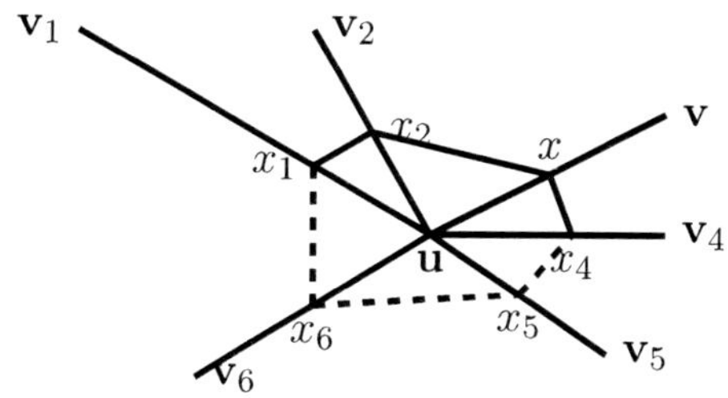


**Fig. 1** Movement of interior junction

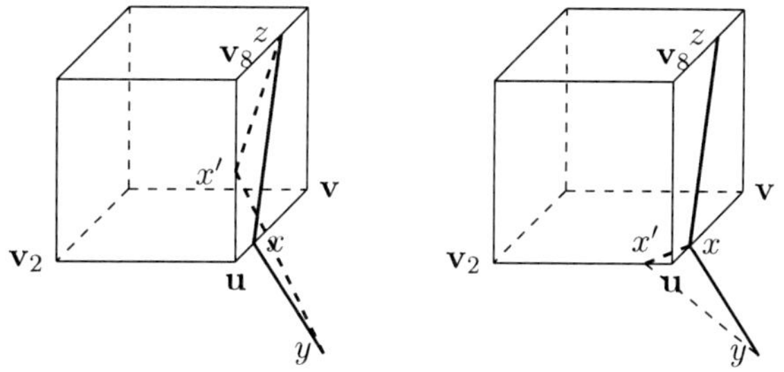


**Fig. 2** Movement of exterior junction $x$. The left figure corresponds to case (i) and the right corresponds to case (ii)

### 2.4.4 Add and Remove Junctions

During the evolution of each point, we may need to add or eliminate junction points. When two neighboring junctions $x$, $y$ are both exterior and $\overline{xy}$ intersects with obstacle $P_{k_1}, P_{k_2}, \cdots, P_{k_r}$ after evolution, we initialize a path with $x$, $y$ being the starting and ending points and $\{P_{k_i}\}_{i=1}^{r}$ being the obstacles. Denote the new added junctions by $(x_{n+1}, x_{n+2}, \cdots, x_{n+s})$ where $s$ is the total number of new junctions. Then they are inserted into the set of junctions in order and the evolution process continues. On the other hand, when two neighboring junctions $x$, $y$ are both exterior and $x$ meets $y$, we may shorten the path by removing $x$ and $y$. More precisely, let $z_1$ be the other neighbor of $x$ and $z_2$ be the other neighbor of $y$, i.e. the path contains $(\cdots, z_1, x, y, z_2, \cdots)$ as a fraction. Since $x = y$, we may connect $z_1$ and $z_2$ directly which shortens the length. In other words, we have the new fraction $(\cdots, z_1, z_2, \cdots)$. Notice, the line segment $\overline{z_1 z_2}$ may intersect with some obstacles. Again we add the necessary junctions as described above. The determination of whether a line segment $\overline{xy}$ intersects with a face can be done by checking whether the intersection point of the line containing $\overline{xy}$ and the surface containing the face lies on the face or not.

## 2.5 Algorithm

We present our algorithm below

**Input**: number of intermittent diffusion intervals $m$, duration of diffusion $\Delta T_l = T_l - S_l, l \leq m$. diffusion coefficients $\sigma_l, l \leq m$.
**Output**: The optimal set $U_{opt}$ of junctions.

1 Initialization. Find the initial set $U$ of junction points.
2 **for** $l = 1 : m$
3 $\quad U_l = U$;
4 $\quad$ **for** $x_i = (\mathbf{u}, \mathbf{v}, \theta_i^0) \in U_l$
5 $\quad\quad$ **for** $j = 1 : \Delta T_l$
6 $\quad\quad\quad$ Update $x$ according to (7), i.e. $\theta_i^{j+1} = \theta_i^j + (-\nabla J(\theta_i^j) + \sigma_l \sqrt{\Delta t} \xi) \Delta t$;
7 $\quad\quad\quad$ Update set $U_l$, i.e. remove junctions from or add junctions to $U_l$;
8 $\quad\quad$ **end**
9 $\quad\quad$ **while** $|\theta_{i+1}^{j+1} - \theta_i^j| > \epsilon$ (or other convergence criterion)
10 $\quad\quad\quad$ Update $x$ according to (6), i.e. $\theta_i^{j+1} = \theta_i^j - \nabla J(\theta_i^j) \Delta t$;
11 $\quad\quad\quad$ Update set $U_l$;
12 $\quad\quad$ **end**
13 $\quad$ **end**
14 **end**
15 Compare $U_l$s and set $U_{opt} = \text{argmin}_{l \leq m} \mathcal{L}(U_l)$.

## 2.6 Complexity Analysis

We now give a brief analysis of the algorithm. Following [16], instead of discussing the algebraic complexity of the algorithm, we will consider the running time in order to achieve certain relative error $\epsilon$.

(1) The initialization is done by constructing the visibility graph and Dijkstra's algorithm. Constructing the visibility graph takes $O(|V|^2)$ while Dijkstra's algorithm takes $O(|E| + |V| \log |V|)$. These two steps are exact in the sense that the complexities do not depend on $\epsilon$. Therefore, they are not counted in the final complexity.
(2) Inner loop line 5–8 takes $O(\Delta T_l)$ time. This is because Eq. (7) takes constant time, and so does adding or removing junctions.
(3) Inner loop line 9-12 takes $T(\epsilon)$ time where $T(\epsilon)$ denotes the number of iterations required until the error is less than $\epsilon$. If we assume the Hessian matrix of the gradient is nondegenerate, which is the case for all polyhedral obstacles [3], then $T(\epsilon) = O(\log \frac{1}{\epsilon})$.

**Table 1** Complexity comparison to other Algorithms

| Algorithm | Complexity |
|---|---|
| $A^*$ | $O((\frac{1}{\epsilon})^3 \log \frac{1}{\epsilon})$ |
| Papadimitriou [16] | $O(\frac{1}{\epsilon})$ |
| Choi et. al. [3] (When the shortest path is not unique.) | $O(\frac{1}{\epsilon})$ |
| Choi et. al. [3] (When the shortest path is unique.) | $O(\log \frac{1}{\epsilon})$ |

Let $\Delta T = \max_{i \le l} \Delta T_i$. Then the total running time is $O(m(\Delta T + \log \frac{1}{\epsilon}))$. From [4], it can be shown that in order to obtain the desired successful probability $1 - \delta$, the number of realizations must be of order $O(\log \frac{1}{\delta})$. Therefore, the complexity is $O(\log \frac{1}{\delta} \log \frac{1}{\epsilon})$. Table 1 shows a complexity comparison with some existing methods.

## 3 Numerical Examples

We show several examples in this section to illustrate the paths obtained by our algorithm. The diffusion coefficients are chosen randomly in interval $[1, 2]$ and the duration of diffusion $\Delta T_l$ is chosen randomly in $[5, 20]$. The parameter $m$, the number of intermittent diffusion intervals $\{[S_l, T_l]\}_1^m$, on which the random perturbations are added to the process, are specified in each example.

**Example 1** The first example computes the shortest path between two points on a hexagonal prism with side length $\sqrt{3}$ and base length 1. In one realization with $m = 10$ intermittent diffusion intervals, it finds 3 minimizers among which the global one, as illustrated in the left plot in Fig. 3, is visited 6 times. In this example, one can easily enumerate all possible combinations to conclude the path obtained with length $L = 2.6$ is the global optimal solution.

(a) Occurs 6 times, L=2.600

(b) Occurs 3 times, L=2.623

(c) Occurs once, L=3.000

**Fig. 3** Example 1

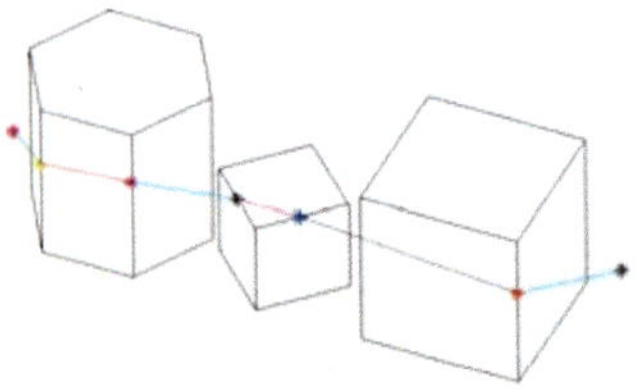

(a) Occurs 13 times, L=7.0803

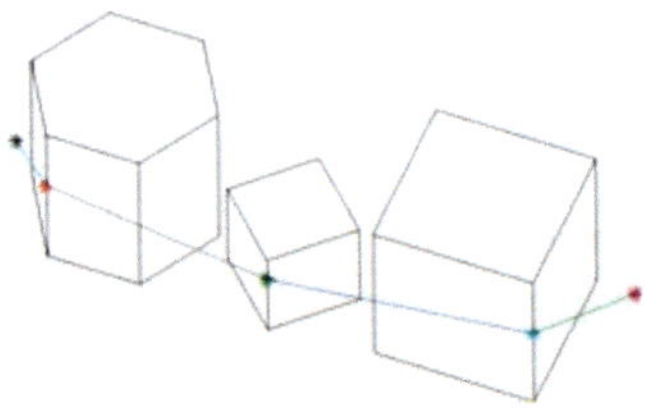

(b) Occurs twice, L=7.0956

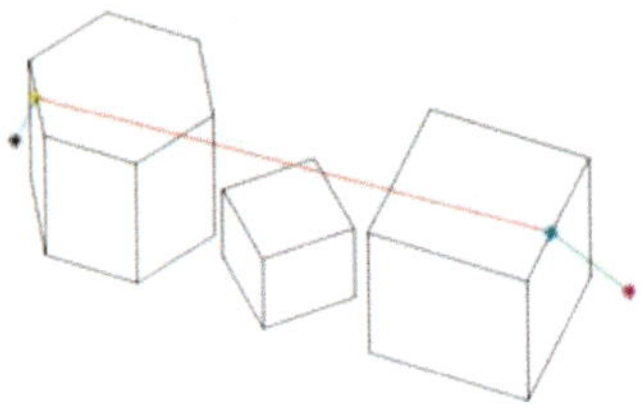

(c) Occurs twice, L=7.3404

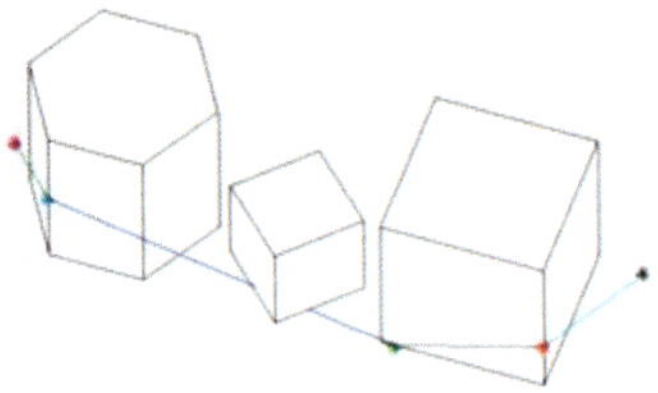

(d) Occurs once, L=7.3436

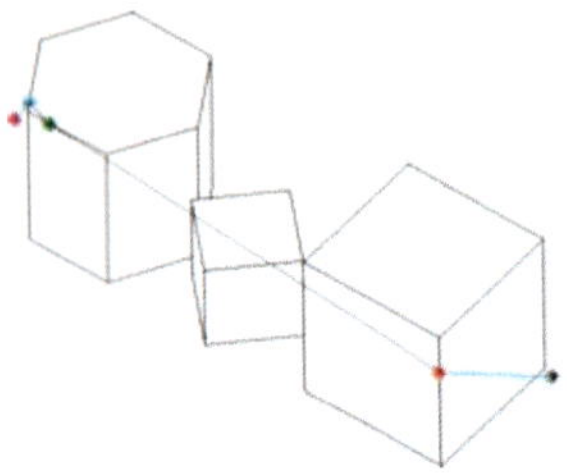

(e) Occurs once, L=7.4314

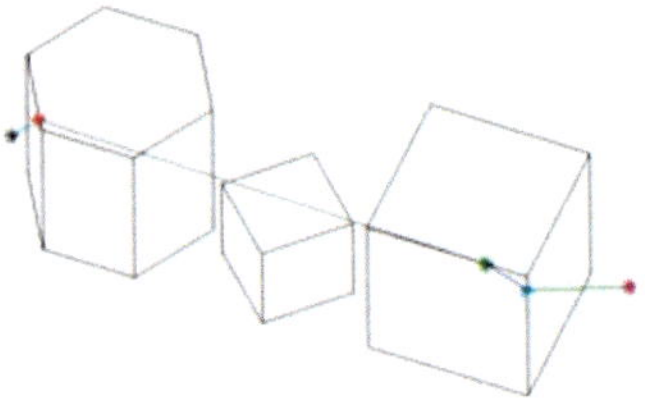

(f) Occurs once, L=7.5253

**Fig. 4** Example 2

**Example 2** There are three obstacles in this example (Fig. 4), two cubes and one hexagonal prism. The algorithm finds 6 local optimal paths in 20 intermittent diffusion intervals, among which the global optimal path occurs 13 times. Below we list all the local minimizers.

**Example 3** In this example (Fig. 5), we demonstrate that our algorithm works for non-convex obstacles without holes. One obstacle is a rotated cube and the other one is a larger cube with an unpenetrated indentation. In 20 intermittent diffusion intervals, the algorithm finds 4 locally optimal path. The globally shortest path is visited 14 times.

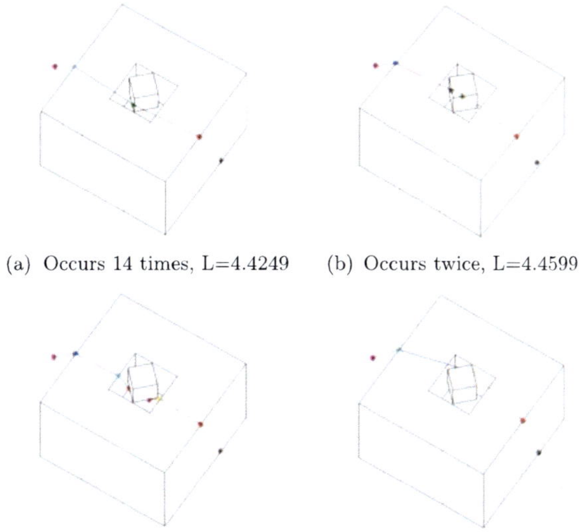

(a) Occurs 14 times, L=4.4249 (b) Occurs twice, L=4.4599

(c) Occurs 3 times, L=4.6176 (d) Occurs once, L=4.5509

**Fig. 5** Example 3

# 4 Polyhedron with Holes

We say two paths are homotopic if one can be deformed continuously to the other while keeping its endpoints fixed. More precisely, let $\mathcal{X}$ be the space that takes away all the obstacles, i.e.

$$\mathcal{X} = \mathbf{R}^3 \setminus \bigcup P_i.$$

Two paths $f_0, f_1$ are path-homotopic if there exists a family of paths $f_t : [0, 1] \to \mathcal{X}$ such that

(1) $f_t(0) = x_0$ and $f_t(1) = x_1$ are fixed.
(2) the map $F : [0, 1] \times [0, 1] \to \mathcal{X}$ given by $F(s, t) = f_t(s)$ is continuous.

Intuitively, two paths are homotopic if one can be continuously transformed to the other without passing through the obstacles. Path-homotopy is an equivalence relation. Thus one can divide all paths into equivalence classes. It is easy to see the following

**Theorem 2** *If all the obstacles have Euler characteristic 2, then there is only one path-homotopy equivalence class in the set of feasible paths $F$.*

***Proof*** Since each $P_i$ has Euler characteristic 2, $P_i$ is homotopic to 2-dimensional sphere $S^2$. Notice that $R^3 - B^3$ where $B^3$ is the 3-dimensional ball is simple-connected. Therefore, any two path in $R^3 - B^3$ are homotopic [8]. Same result holds for $R^3$ taking away $n$ balls.

With this result, it is simple to obtain the results in Theorem 1.

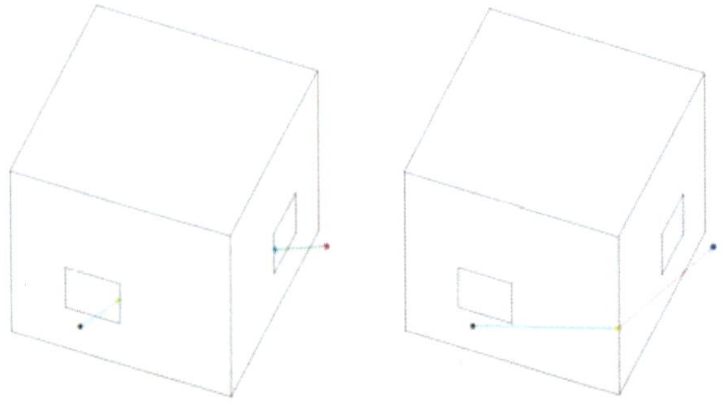

(a) L=1.1543 (b) L=1.2659

**Fig. 6** Shortest path with tunneled cube

*Proof of Theorem* 1 Theorem 2 guarantees that our algorithm is able to visit all possible paths from any initialization. The rest of the theorem is simply the statement from [4] and hence omitted.

However, on the contrary, if the obstacle contains holes, for example, a triangulated torus, there would be multiple equivalence classes. For illustration, see the following tunneled cube. The shortest path through the hole is 1.1543 while the one that doesn't penetrate the hole has length 1.2659. By slightly changing the position of the hole, the shortest path would be the one that does not pass through it. Therefore, multiple initializations are needed to ensure that all possible equivalence classes are covered.

A simple idea we can use is to "block" the homotopy equivalence class the current path belongs to and then reinitialize. "Blocking" means deleting some vertices of the obstacles such that the reinitialization will force the new path to a different homotopy class. After the gradient descent settles down at the global minimizer in the current homotopy class, the path is reinitialized and the algorithm is repeated to get a different global minimizer. This procedure is repeated until all homotopy equivalence classes are visited. The two paths in the above example are obtained by this method (Fig. 6).

However, there are two problems with this approach. First of all, the block is often difficult to form because which vertices should be removed is a complicated matter, for instance, a well triangulated torus as follows (Fig. 7). Second, the number of different homotopy classes we need to visit is unknown in advance. For example, topologically, there are infinitely many homotopy classes for a smooth torus and the shortest path could wind the torus arbitrary times. In Fig. 8, the shortest path winds the torus twice.

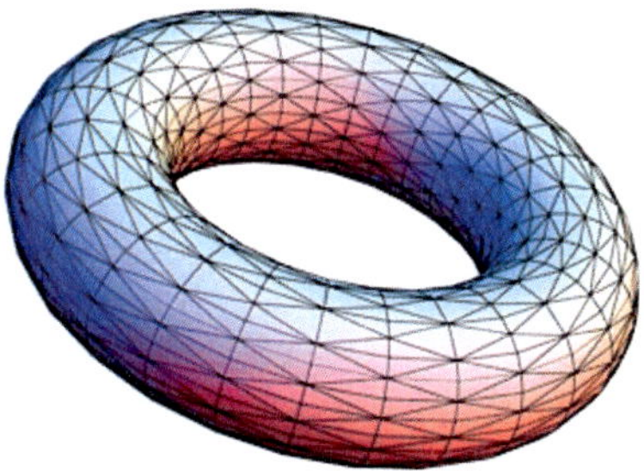

**Fig. 7** A triangulated torus

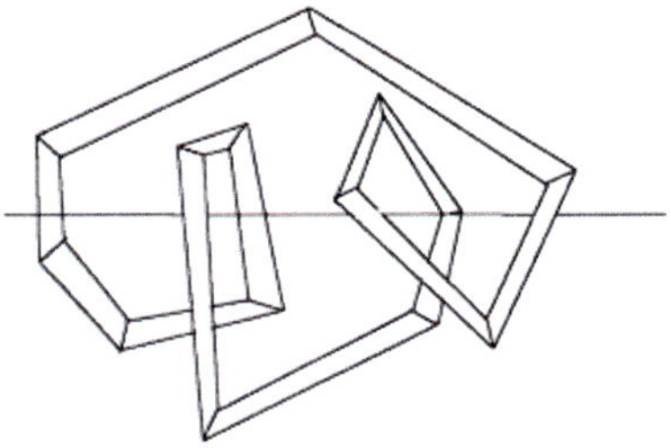

**Fig. 8** A shortest path winding a torus twice

A different approach is to use an already established approximation method, for example [15] as described in the introduction section, to initialize the path. Those algorithms are able to obtain a path that has length $1+\epsilon$ times the length of the shortest path. Here $\epsilon$ depends on the mesh size. If the mesh size is sufficiently small, the initialized path and the global minimizer will be in the same homotopy class. However, the choice of the grid size is a critical and often hard to determine.

As discussed above, our method still applies for polyhedrons with holes provided that appropriate initializations are taken. Although initialization is a complicated matter, simple ideas usually work for most cases. We conclude our discussion here and leave the improvement of initialization methods to our future work.

## 5 Future Work

The method we propose in this work can be equally applied to a general class of problems. In detail, consider the following problem

$$\gamma_{opt} = \text{argmin}_{\gamma \in \mathbf{F}} L(\gamma). \tag{9}$$

Here $L$ is a general functional on $\mathbf{F}$ with the form

$$L(\gamma) = \int_0^1 l(|\dot{\gamma}(\theta)|)\, \mathrm{d}\theta \tag{10}$$

where $l(x)$ is a convex function. The Euclidean length of the path, which we consider in this paper, simple corresponds to the case where $l(x) = x$. It turns out that in this general setting, the optimal path has the same simple structure as mentioned in this paper. Therefore, the method can be applied without any essential modification. An example is $l(x) = x^2$, in which case the functional $L$ represents the oil consumption of a car. We leave the direction for future exploration.

## References


1. A. Arnold, P. Markowich, G. Toscani, A. Unterreiter, On convex sobolev inequalities and the rate of convergence to equilibrium for fokker-planck type equations (2001)
2. J. Canny, J. Reif, New lower bound techniques for robot motion planning problems, in *28th Annual Symposium on Foundations of Computer Science*, pp. 49–60. IEEE (1987)
3. J. Choi, J. Sellen, C.-K. Yap, Precision-sensitive euclidean shortest path in 3-space, in *Proceedings of the Eleventh Annual Symposium on Computational Geometry*, pp. 350–359. ACM (1995)
4. S.-N. Chow, T.-S. Yang, H. Zhou, Global optimizations by intermittent diffusion. *National Science Council Tunghai University Endowment Fund for Academic Advancement Mathematics Research Promotion Center*, p. 121 (2009)
5. S.-N. Chow, J. Lu, H.M. Zhou, Finding the shortest path by evolving junctions on obstacle boundaries (E-JOB): An initial value ODE's approach. J. Appl Comput Harmon Anal. **35**(1), 156–176 (2013)
6. F. Fahimi, *Autonomous Robots: Modeling, Path Planning, and Control* (Springer, 2008)
7. L.P. Gewali, S. Ntafos, I.G. Tollis, Path planning in the presence of vertical obstacles. IEEE Trans. Robot. Autom. **6**(3), 331–341 (1990)
8. A. Hatcher, *Algebraic Topology* (2002)
9. J. Hershberger, S. Suri, An optimal algorithm for Euclidean shortest paths in the plane. SIAM J. Comput. **28**(6), 2215–2256 (1999)
10. R. Holley, D. Stroock, Logarithmic sobolev inequalities and stochastic ising models. J. Statist. Phys. **46**(5), 1159–1194 (1987)
11. K. Jiang, L.S. Seneviratne, S.W.E. Earles, Finding the 3d shortest path with visibility graph and minimum potential energy, in *Intelligent Robots and Systems' 93, IROS'93. Proceedings of the 1993 IEEE/RSJ International Conference on*, vol. 1, pp. 679–684. IEEE (1993)
12. S. M. LaValle. *Planning Algorithms* (Cambridge University Press, 2006)
13. P.A. Markowich, C. Villani, On the trend to equilibrium for the fokker-planck quation: an interplay between physics and functional analysis. Mat. Contemp. **19**, 1–29 (2000)
14. J.S.B. Mitchell, Shortest path and networks. *Handbook of Discrete and Computational Geometry*, pp. 755–778 (1997)
15. C.H. Papadimitriou, An algorithm for shortest-path motion in three dimensions. Inf. Process. Lett. **20**(5), 259–263 (1985)
16. C.H. Papadimitriou, An algorithm for shortest-path motion in three dimensions. Inf. Process. Lett. **20**(5), 259–263 (1985)
17. J.H. Reif, J.A. Storer, Shortest paths in euclidean space with polyhedral obstacles. Technical report, DTIC Document (1985)
18. J.A. Sethian, A fast marching level set method for monotonically advancing fronts. Proc. Natl Acad. Sci. **93**(4), 1591 (1996)
19. A. Sharir, A. Baltsan, On shortest paths amidst convex polyhedra, in *Proceedings of the Second Annual Symposium on Computational Geometry*, pp. 193–206. ACM (1986)
20. H. Zhao, A fast sweeping method for eikonal equations. Math. Computat. **74**(250), 603–628 (2005)

# Multigrid Methods for Image Registration Model Based on Optimal Mass Transport

**Yangang Chen and Justin W. L. Wan**

**Abstract** In this survey, we present fast, accurate and convergent numerical methods for solving non-rigid image registration based on optimal mass transport. To solve the model equation, we first transform the nonlinear PDEs into an HJB equation. We apply a mixed standard 7-point stencil and semi-Lagrangian wide stencil discretization, such that the numerical solution is guaranteed to converge to the viscosity solution of the Monge-Ampere equation. We design a numerical scheme that converges to the optimal transformation between the target and template images. Finally, we introduce fast multigrid methods for solving the discrete nonlinear system. In particular, we use a four-directional alternating line relaxation scheme as smoother, a coarsening strategy where wide stencil points are set as coarse grid points. Linear interpolation and injection are used in prolongation and restriction, respectively. Our numerical results show that the numerical solution yield good quality transformations for non-rigid image registration and the convergence rates of the proposed multigrid methods are mesh-independent.

**Keywords** Image registration · Optimal mass transport · Monge-Ampere equation · Multigrid · Monotone discretization scheme

## 1 Introduction

In many applications, one has to compare two images $T$ (template) and $R$ (reference) which display the same object, but the object inside the images is not spatially aligned,

Y. Chen
Department of Applied Mathematics, University of Waterloo, 200 University Avenue West, Waterloo, ON, Canada
e-mail: y493chen@uwaterloo.ca

J. W. L. Wan (✉)
David R. Cheriton School of Computer Science, University of Waterloo, 200 University Avenue West, Waterloo, ON, Canada
e-mail: justin.wan@uwaterloo.ca



X.-C. Tai et al. (eds.), *Mathematical Methods in Image Processing and Inverse Problems*, Springer Proceedings in Mathematics & Statistics 360,
https://doi.org/10.1007/978-981-16-2701-9_11

or the devices that record the two images are different. The image registration problem is to find a coordinate transformation $\phi$ which transforms the image $T$ to another image $T_\phi$, such that $T_\phi$ is similar and thus comparable to the image $R$.

One important application of image registration is to compare medical images of the same patient, such as CT (computed tomography) and MRI (magnetic resonance imaging) images of a damaged brain, which gives guidance for diagnosis and surgery [1, 24]. Image registration can also be used for image fusion [26]. Multiple images of the same object are taken, registered and then merged together, such that the integrated image provides more useful than the original ones. We refer readers to [2] for more discussion on applications.

Different approaches have been developed for image registration problems, including parametrized transformation [30, 44], landmark-based registration [36], elastic registration [8], fluid registration [14], diffusion registration [18], demon's registration [41], flow of diffeomorphism [16, 43], etc. A substantial discussion of existing methods can be found in [32, 40].

This paper surveys the three recent works of the authors [10–12] and it considers a non-rigid image registration method based on Monge-Kantorovich mass transport [9, 20, 22, 23, 33, 37]. Optimal mass transport problems appear in many applications and have been widely studied (see e.g. [13, 34, 38]). The use of optimal mass transport for image registration was first proposed in [22, 23]. This image registration model treats two images $R$ and $T$ as two mass densities. The goal is to find a mapping which transforms one mass density $T$ to the other $R$ with mass conservation. Such transformation is non-unique. By defining a transformation-dependent cost function and minimizing it, we can obtain a unique optimal transformation. This optimal transformation has desirable properties. For instance, it is usually diffeomorphic and does not introduce foldings and crossings.

The primary advantage of this image registration model is that, unlike many other non-rigid methods that are only applicable for images with small deformations, this model can be applied to images with large deformations. See Figs. 1 and 2 in [22] for an example of images with large deformations. Indeed, given *any* $R$ and $T$, the transformed image $T_\phi$ under the mass transport formulation can be equal to $R$ [33].

Numerical methods have been developed for solving the image registration model based on optimal mass transport. In [22, 23], the authors construct an initial mass-preserving mapping $\phi_0$ by solving a nonlinear partial differential equation (PDE), and obtain a second mass-preserving mapping $\phi_s$ by solving another nonlinear PDE system, such that $\phi_0 \circ \phi_s$ is the optimal transformation. The entire process involves many intermediate steps. Also, in general, a nonlinear PDE (or PDE system) has multiple solutions. An immediate challenge is that the nonlinear PDE system in [22, 23] can give multiple transformations between $R$ and $T$, which may not be the optimal transformation.

An alternative approach is to solve an equivalent nonlinear Monge-Ampère equation. The gradient of the unique globally convex solution corresponds to the optimal transformation between $R$ and $T$ [22, 27]. The convex solution itself is usually called a scalar potential that generates the optimal transformation. Some literature has investigated numerical schemes for the Monge-Ampère equation arising from the image registration model [9, 20, 37]. However, for the approach in [20], the computational cost per pixel must increase to infinity as the image size increases [17]. The methods in [9, 37] are based on gradient descent, which is essentially equivalent to solving the Monge-Ampère equation using explicit pseudo-timestepping.

In this survey paper, we present a numerical approach for the image registration model based on optimal mass transport by solving the equivalent Monge-Ampère equation. In order to ensure that the numerical scheme yields the optimal transformation between $R$ and $T$ [20, 21], we will adopt a monotone finite difference discretization method based on our previous work [10], which can be proved that the resulting numerical solution converges to the viscosity solution [4] of the Monge-Ampère equation. We will also present efficient multigrid methods for solving the resulting nonlinear discretized system [11].

Standard multigrid methods turn out to have poor convergence. There are two major factors behind the poor convergence. One is that the PDE may become anisotropic along various directions. Standard pointwise smoothers fail to smooth the error along the weakly connected directions. The other factor is that the resulting matrix is non-symmetric, which is a well-known issue when applying multigrid. Algebraic multigrid (AMG) methods [35, 39] have been used as preconditioners. However, they are not efficient as stand-alone solvers since AMG methods assess geometric information indirectly though the strength of connections which is not effective for the monotone discretization.

To obtain a fast stand-alone multigrid solver for solving Monge-Ampère equations, we note that wide stencils introduce oscillations locally to the error, and such oscillations cannot be eliminated by smoothers, including the alternating line smoothers. However, the oscillations are restricted at the wide stencil points. One possible solution to capture the oscillations is to use a sophisticated interpolation, which can be complicated and expensive to set up. Instead, we use a non-standard coarsening strategy. Specifically, we set wide stencil points as coarse grid points. The purpose is to directly use the coarse grid points to capture the oscillations. As the wide stencils are mainly restricted to the singular points or singular lines, setting wide stencil points as coarse grid points does not significantly increase the number of the coarse grid points. In our numerical experiments, we illustrate that the proposed multigrid method has a mesh-independent convergence rate for various problems.

This paper is organized as follows. In Sect. 2, we describe the image registration model based on optimal mass transport. Section 3 describes a finite difference discretization for the Monge-Ampère equation arising from the mass transport image registration model. In Sect. 4, we present efficient multigrid methods to solve the discretized system. Numerical results in Sect. 5 show that our multigrid methods converge quickly with mesh-independent convergence rate. Image results are also provided to demonstrate the performance of the registration model. Section 6 concludes the paper.

## 2 Image Registration Model Based on Optimal Mass Transport

### 2.1 Image Registration

Given a template image $T$ and a reference image $R$, the objective of the image registration problem is to align the two images. Mathematically, we consider the template (reference) image as a function defined on the domain $\Omega^T$ ($\Omega^R$). For simplicity, we assume that $\Omega^T = \Omega^R = [0, 1] \times [0, 1]$. The image registration problem can be formulated as to find a coordinate transformation $\phi$ that minimizes the difference between $\rho^{T_\phi}$ and $\rho^R$, where $\rho^T$ and $\rho^R$ are the intensities of the template image and reference image, respectively, and $\rho^{T_\phi}$ is the intensity of the transformed image, $T_\phi$. The intensities must be positive and bounded. The difference of the two images is usually measured by some function such as sum of squared differences

$$\mathcal{D}(\rho^{T_\phi}, \rho^R) \equiv \|\rho^{T_\phi} - \rho^R\|_{L_2(\Omega^R)}. \tag{1}$$

### 2.2 Optimal Mass Transport Model

Consider registering two images $T$ and $R$. If we view them as two piles of soil with the densities $\rho^T$ and $\rho^R$, then an image registration problem can be interpreted as a mass transport problem [22, 23, 33]. That is, we consider two piles of soil $\rho^T$ and $\rho^R$ with the same total mass:

$$\int_{\hat{x}\in\Omega^T} \rho^T(\hat{x})d^2\hat{x} = \int_{x\in\Omega^R} \rho^R(x)d^2x. \tag{2}$$

The image registration problem becomes to find a coordinate transformation $\phi : \Omega^R \to \Omega^T$, or $\hat{x} = \phi(x) \in \mathbb{R}^2$, such that $\rho^T$ is transformed to $\rho^R$ while the total mass is conserved:

$$\int_{x\in\Omega^R} \rho^T(\phi(x))d^2\phi(x) = \int_{x\in\Omega^R} \rho^R(x)d^2x, \tag{3}$$

or equivalently,

$$\rho^T(\phi(x)) \det[D\phi(x)] = \rho^R(x), \tag{4}$$

where $D\phi(x) \in \mathbb{R}^{2\times2}$ is the Jacobian of the transformation $\phi(x)$.

Under the transformation $\phi$, define the intensity of the transformed image $T_\phi$ as

$$\rho^{T_\phi}(x) \equiv \rho^T(\phi(x)) \det[D\phi(x)]. \tag{5}$$

Then the transformed template image is equal to the reference image:

$$\rho^{T_\phi}(\boldsymbol{x}) = \rho^R(\boldsymbol{x}). \tag{6}$$

As a result, the mass transport model can transform any template image $T$ to any reference image $R$ [33].

The mass transport registration (5) is ill-posed. More specifically, there exist multiple transformations that move the soil $\rho^T$ to $\rho^R$. Among all possible transformations, one of them requires the "least cost", which is desirable. Following [5, 22, 23], we aim to find the optimal transformation $\phi^*(\boldsymbol{x})$ that minimizes the following cost function:

$$\phi^*(\boldsymbol{x}) \equiv \underset{\phi(\boldsymbol{x})}{\arg\min} \int_{\mathbb{R}^2} \|\boldsymbol{x} - \phi(\boldsymbol{x})\|^2 \rho^R(\boldsymbol{x}) d^2\boldsymbol{x}, \tag{7}$$

which is the weighted least squares displacement of the mass. In essence, (7) regularizes the mass transport registration and makes the transformation between $\rho^T$ and $\rho^R$ unique.

## 2.3 *Monge-Ampère Equation*

It has been proved in [27] that the optimal transformation that minimizes the cost function (7) can be written as

$$\phi^*(\boldsymbol{x}) = \nabla u(\boldsymbol{x}), \tag{8}$$

where $u \in C(\Omega^R)$ is a strictly convex scalar potential field, and its gradient $\nabla u$ generates the optimal transformation $\phi^*$. Substituting (8) into (4), we have

$$\det[D^2 u(\boldsymbol{x})] = \frac{\rho^R(\boldsymbol{x})}{\rho^T(\nabla u(\boldsymbol{x}))}, \tag{9}$$

$$u \text{ is strictly convex.} \tag{10}$$

Equations (9)–(10) is a **Monge-Ampère equation**.

Due to the nonlinearity, the equation (9) itself, without the convexity constraint (10), can have multiple solutions [6, 17]. However, the solution of (9) that satisfies the convexity constraint (10) is unique [20], which we will denote as $u^*$ whenever we need to distinguish it from the other solutions. We emphasize that the convexity of $u^*$ is equivalent to the optimality of the transformation $\phi^* = \nabla u^*$ [20, 22].

## 3 Finite Difference Discretization

In order to design a finite difference scheme that converges to the viscosity solution, we first convert the Monge-Ampère equation into an equivalent Hamilton-Jacobi-Bellman (HJB) equation. The equivalence of the two PDEs is first established in [28, 29]. Here we present the equivalent HJB equation as follows:

**Theorem 1** *Let $u \in C^2(\Omega^R)$ be convex, and let $\rho^T \in C(\Omega^T)$ and $\rho^R \in C(\Omega^R)$ be two positive functions. Then the Monge-Ampère equation (9)–(10) is equivalent to the following HJB equation*

$$\hat{\mathcal{L}}_{c^*(\boldsymbol{x}),\theta^*(\boldsymbol{x})}\, u(\boldsymbol{x}) = 0, \tag{11}$$

$$\text{subject to } (c^*(\boldsymbol{x}), \theta^*(\boldsymbol{x})) \equiv \mathop{\arg\max}_{(c(\boldsymbol{x}),\theta(\boldsymbol{x}))\in\Gamma} \hat{\mathcal{L}}_{c(\boldsymbol{x}),\theta(\boldsymbol{x})}\, u(\boldsymbol{x}), \tag{12}$$

*where the differential operator is*

$$\begin{aligned}\hat{\mathcal{L}}_{c(\boldsymbol{x}),\theta(\boldsymbol{x})}\, u(\boldsymbol{x}) \equiv {} & -\sigma_{11}(c(\boldsymbol{x}),\theta(\boldsymbol{x}))u_{xx}(\boldsymbol{x}) - 2\sigma_{12}(c(\boldsymbol{x}),\theta(\boldsymbol{x}))u_{xy}(\boldsymbol{x}) \\ & -\sigma_{22}(c(\boldsymbol{x}),\theta(\boldsymbol{x}))u_{yy}(\boldsymbol{x}) + 2\sqrt{c(\boldsymbol{x})(1-c(\boldsymbol{x}))\frac{\rho^R(\boldsymbol{x})}{\rho^T(\nabla u(\boldsymbol{x}))}},\end{aligned} \tag{13}$$

*and $(c(\boldsymbol{x}), \theta(\boldsymbol{x}))$ is the pair of control at point $\boldsymbol{x}$, $\Gamma = [0, 1] \times \left[-\frac{\pi}{4}, \frac{\pi}{4}\right)$ is the set of admissible control. The coefficients are*

$$\begin{aligned}\sigma_{11}(c(\boldsymbol{x}),\theta(\boldsymbol{x})) &= \frac{1}{2}[1-(1-2c(\boldsymbol{x}))\cos 2\theta(\boldsymbol{x})], \\ \sigma_{22}(c(\boldsymbol{x}),\theta(\boldsymbol{x})) &= \frac{1}{2}[1+(1-2c(\boldsymbol{x}))\cos 2\theta(\boldsymbol{x})], \\ \sigma_{12}(c(\boldsymbol{x}),\theta(\boldsymbol{x})) &= \frac{1}{2}(1-2c(\boldsymbol{x}))\sin 2\theta(\boldsymbol{x}).\end{aligned} \tag{14}$$

Below, we will describe a monotone finite difference discretization scheme for the HJB equation (11)–(12).

### *3.1 Standard 7-Point Stencil Discretization*

Consider discretizing the differential operator (13) at a grid point $\boldsymbol{x}_{i,j}$. We use the standard central differencing to approximate $u_{xx}(\boldsymbol{x}_{i,j})$ and $u_{yy}(\boldsymbol{x}_{i,j})$. Regarding the cross derivative $u_{xy}(\boldsymbol{x}_{i,j})$, it can be shown that the standard 7-point stencil discretization leads to a monotone scheme in the following two cases:

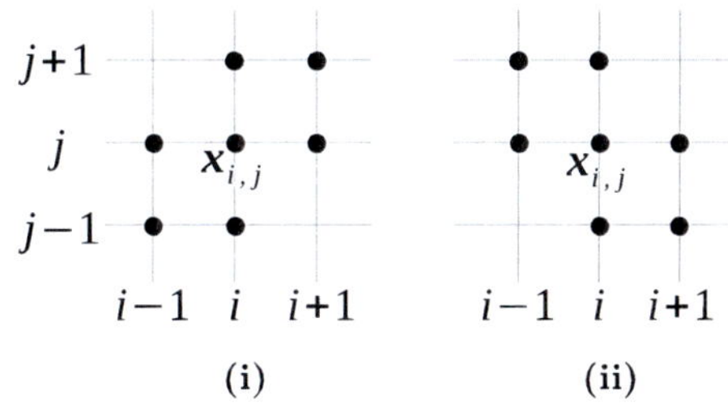


**Fig. 1** **(i)** 7-point stencil of (16); **(ii)** 7-point stencil of (18)

- **Case 1.** When the coefficients (14) at a grid point $\boldsymbol{x}_{i,j}$ satisfy

$$\sigma_{11}(c_{i,j},\theta_{i,j}) \geq |\sigma_{12}(c_{i,j},\theta_{i,j})|,\ \sigma_{22}(c_{i,j},\theta_{i,j}) \geq |\sigma_{12}(c_{i,j},\theta_{i,j})|,\ \sigma_{12}(c_{i,j},\theta_{i,j}) \geq 0, \tag{15}$$

we approximate $u_{xy}(\boldsymbol{x}_{i,j})$ using

$$\frac{1}{2}(D_x^+D_y^+ + D_x^-D_y^-)u_{i,j} \equiv \frac{2u_{i,j}+u_{i+1,j+1}+u_{i-1,j-1}-u_{i+1,j}-u_{i-1,j}-u_{i,j+1}-u_{i,j-1}}{2h^2}. \tag{16}$$

- **Case 2.** When the coefficients (14) at a grid point $\boldsymbol{x}_{i,j}$ satisfy

$$\sigma_{11}(c_{i,j},\theta_{i,j}) \geq |\sigma_{12}(c_{i,j},\theta_{i,j})|,\ \sigma_{22}(c_{i,j},\theta_{i,j}) \geq |\sigma_{12}(c_{i,j},\theta_{i,j})|,\ \sigma_{12}(c_{i,j},\theta_{i,j}) \leq 0, \tag{17}$$

we approximate $u_{xy}(\boldsymbol{x}_{i,j})$ using

$$\frac{1}{2}(D_x^+D_y^- + D_x^-D_y^+)u_{i,j} \equiv \frac{-2u_{i,j}-u_{i+1,j-1}-u_{i-1,j+1}+u_{i+1,j}+u_{i-1,j}+u_{i,j+1}+u_{i,j-1}}{2h^2}. \tag{18}$$

Figure 1 shows the stencil points of the 7-point stencil discretizations (16) and (18).
As a result, the discretization of the differential operator (13) at $\boldsymbol{x}_{i,j}$ reads

$$\begin{aligned}\mathcal{L}_{i,j}(c_{i,j},\theta_{i,j};u_h) \equiv &-\sigma_{11}(c_{i,j},\theta_{i,j})D_x^+D_x^-u_{i,j} - \sigma_{12}(c_{i,j},\theta_{i,j})(D_x^+D_y^{\pm} + D_x^-D_y^{\mp})u_{i,j}\\ &-\sigma_{22}(c_{i,j},\theta_{i,j})D_y^+D_y^-u_{i,j} + 2\sqrt{c_{i,j}(1-c_{i,j})}f_{i,j}.\end{aligned} \tag{19}$$

## 3.2 *Semi-Lagrangian Wide Stencil Discretization*

However, if neither of Conditions (15) and (17) is fulfilled at the grid point $\boldsymbol{x}_{i,j}$, then it is unclear how to directly discretize the cross derivative $u_{xy}(\boldsymbol{x}_{i,j})$ in (13) monotonically. Our approach is to consider a **semi-Lagrangian wide stencil discretization** [15, 31]. Figure 2 illustrates the discretization process. More specifically, we consider eliminating the cross derivative $u_{xy}(\boldsymbol{x}_{i,j})$ by a local coordinate transformation. Let $\{(\boldsymbol{e}_z)_{i,j}, (\boldsymbol{e}_w)_{i,j}\}$ be a local orthogonal basis obtained by a clockwise rotation

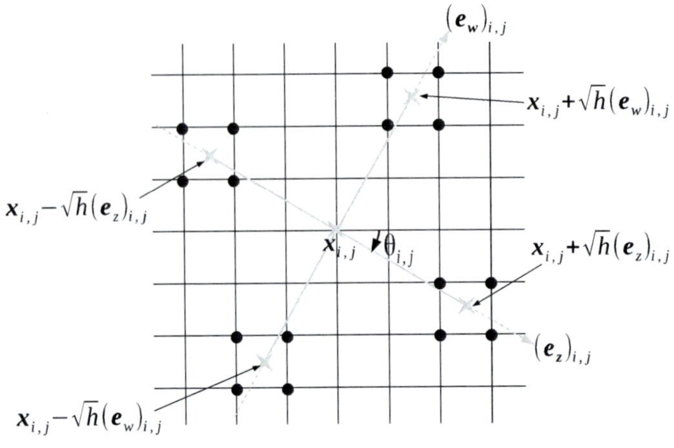


**Fig. 2** Semi-Lagrangian wide stencil discretization at a grid point $\boldsymbol{x}_{i,j}$ inside the computational domain

of the standard axes $\{(\boldsymbol{e}_x)_{i,j}, (\boldsymbol{e}_y)_{i,j}\}$, as represented by the grey axes in Fig. 2. By straightforward algebra, one can show that if the rotation angle is

$$\frac{1}{2}\arctan\frac{2\sigma_{12}\left(c_{i,j},\theta_{i,j}\right)}{\sigma_{22}\left(c_{i,j},\theta_{i,j}\right)-\sigma_{11}\left(c_{i,j},\theta_{i,j}\right)}=\theta_{i,j},$$

then the cross derivative vanishes under the basis $\{(\boldsymbol{e}_z)_{i,j}, (\boldsymbol{e}_w)_{i,j}\}$. As a result, (13) becomes

$$-c_{i,j}\,u_{zz}(\boldsymbol{x}_{i,j})-\left(1-c_{i,j}\right)u_{ww}(\boldsymbol{x}_{i,j})+2\sqrt{c_{i,j}\left(1-c_{i,j}\right)f_{i,j}} \tag{20}$$

Here $u_{zz}(\boldsymbol{x}_{i,j})$ and $u_{ww}(\boldsymbol{x}_{i,j})$ are the directional derivatives along the basis $(\boldsymbol{e}_z)_{i,j}$ and $(\boldsymbol{e}_w)_{i,j}$. We note that (20) still depends on $\theta_{i,j}$, as the basis $(\boldsymbol{e}_z)_{i,j}$ and $(\boldsymbol{e}_w)_{i,j}$ depend on $\theta_{i,j}$.

To discretize (20), one may consider applying the standard central differencing to $u_{zz}(\boldsymbol{x}_{i,j})$ and $u_{ww}(\boldsymbol{x}_{i,j})$. For instance, we approximate $u_{zz}(\boldsymbol{x}_{i,j})$ by

$$\frac{1}{h^2}\left[u(\boldsymbol{x}_{i,j}+h(\boldsymbol{e}_z)_{i,j})-2u_{i,j}+u(\boldsymbol{x}_{i,j}-h(\boldsymbol{e}_z)_{i,j})\right]. \tag{21}$$

However, since the stencil is rotated, the stencil points $\boldsymbol{x}_{i,j}\pm h(\boldsymbol{e}_z)_{i,j}$ may no longer coincide with any grid points. In such cases, we consider approximating $u(\boldsymbol{x}_{i,j}\pm h(\boldsymbol{e}_z)_{i,j})$ using bilinear interpolation from the neighboring grid points. However, a consequence of the bilinear interpolation is that the truncation error of (21) turns out to be $O(1)$, which is not consistent. In order to maintain consistency, we choose the stencil length $\sqrt{h}$, which yields $O(h)$ truncation error. We note that the stencil length $\sqrt{h}$ is greater than $h$, which gives rise to a "wide" stencil.

Under the stencil length $\sqrt{h}$, the new stencil points are $\boldsymbol{x}_{i,j} \pm \sqrt{h}(\boldsymbol{e}_z)_{i,j}$ and $\boldsymbol{x}_{i,j} \pm \sqrt{h}(\boldsymbol{e}_w)_{i,j}$, as represented by the grey stars in Fig. 2. The unknown values at these stencil points are approximated by the bilinear interpolation from their neighboring points, as represented by the black dots in Fig. 2. We denote these interpolated unknown values as $\mathcal{I}_h u|_{\boldsymbol{x}_{i,j}\pm\sqrt{h}(\boldsymbol{e}_z)_{i,j}}$ and $\mathcal{I}_h u|_{\boldsymbol{x}_{i,j}\pm\sqrt{h}(\boldsymbol{e}_w)_{i,j}}$. The finite difference discretizations for $u_{zz}(\boldsymbol{x}_{i,j})$ and $u_{ww}(\boldsymbol{x}_{i,j})$ are then given by

$$D_z^+ D_z^- u_{i,j} \equiv \frac{\mathcal{I}_h u|_{\boldsymbol{x}_{i,j}+\sqrt{h}(\boldsymbol{e}_z)_{i,j}} - 2u_{i,j} + \mathcal{I}_h u|_{\boldsymbol{x}_{i,j}-\sqrt{h}(\boldsymbol{e}_z)_{i,j}}}{h}, \tag{22}$$

$$D_w^+ D_w^- u_{i,j} \equiv \frac{\mathcal{I}_h u|_{\boldsymbol{x}_{i,j}+\sqrt{h}(\boldsymbol{e}_w)_{i,j}} - 2u_{i,j} + \mathcal{I}_h u|_{\boldsymbol{x}_{i,j}-\sqrt{h}(\boldsymbol{e}_w)_{i,j}}}{h}. \tag{23}$$

Finally, the discretization of the differential operator (13) at $\boldsymbol{x}_{i,j}$ reads

$$\mathcal{L}_{i,j}(c_{i,j}, \theta_{i,j}; u_h) \equiv -c_{i,j}\, D_z^+ D_z^- u_{i,j} - \left(1 - c_{i,j}\right)\, D_w^+ D_w^- u_{i,j} + 2\sqrt{c_{i,j}(1 - c_{i,j})f_{i,j}}. \tag{24}$$

We remark that here we have only discussed the scenario where $\boldsymbol{x}_{i,j}$ is well inside the computational domain. The scenario where $\boldsymbol{x}_{i,j}$ is near the boundary can be handled similarly.

### 3.3 *Mixed Discretization*

The advantage of the semi-Lagrangian wide stencil discretization (24) is that it is unconditionally monotone but it is only first order accurate. On the other hand, the standard 7-point stencil discretization is second order accurate. In order to combine the advantages of both discretization schemes, we will only apply the semi-Lagrangian wide stencil discretization at the grid points where neither (15) nor (17) is satisfied. Otherwise, the standard 7-point stencil discretization is applied. The resulting discretization method can be written as:

The significance of this mixed discretization is that monotonicity is strictly maintained at every grid point, and meanwhile, by using the standard 7-point stencil discretization as much as possible, the numerical scheme is as accurate as possible.

The mixed discretization scheme gives rise to a nonlinear discrete system which can be written in the following matrix form:

$$A_h(c_h^*, \theta_h^*)\, u_h = b_h(c_h^*, \theta_h^*), \tag{25}$$

$$\text{subject to} \quad (c_h^*, \theta_h^*) \equiv \underset{(c_h,\theta_h)\in\Gamma}{\arg\max} \left\{A_h(c_h, \theta_h)\, u_h - b_h(c_h, \theta_h)\right\}, \tag{26}$$

where the matrix $A_h \in \mathbb{R}^{n_x n_y \times n_x n_y}$ and the vectors $u_h, c_h, \theta_h, b_h \in \mathbb{R}^{n_x n_y}$.

**Algorithm 1** Mixed discretization for the HJB equation (11)-(12)

**for** $i = 1, , ..., n_x$ **do**
  **for** $j = 1, , ..., n_y$ **do**
    **if** $(c_{i,j}, \theta_{i,j})$ satisfies Conditions (15) or (17) **then**
      The discrete equation at $(i, j)$, $\mathcal{L}_{i,j}(c_{i,j}, \theta_{i,j}; u_h)$, is given by the **standard 7-point stencil discretization** (19)
    **else**
      The discrete equation at $(i, j)$, $\mathcal{L}_{i,j}(c_{i,j}, \theta_{i,j}; u_h)$, is given by the **semi-Lagrangian wide stencil discretization** (24)
    **end if**
  **end for**
**end for**

# 4 Multigrid Methods

We will apply multigrid methods for solving (25). We start with multigrid methods for the standard 7-point stencil discretization. More precisely, we consider the case where the standard 7-point stencil discretization can be applied on the entire computational domain and still results in a monotone scheme. We will leave the discussion of multigrid for more general mixed stencil discretization to Sect. 4.3.

## *4.1 Policy-MG Iteration*

One family of multigrid methods for solving the discretized HJB equation (25) is based on a global Newton-like iteration for the nonlinear system, called policy iteration (or Howard's algorithm) [19, 25]. At each policy iteration, a linear multigrid solver is applied to solve the linearized system. The algorithm can be written as follows:

Start with an initial guess of the solution $u_h^{(0)}$.
For $k = 0, 1, ...$ until convergence:

1. Solve for the optimal control pair $(a_h^{(k)}, \theta_h^{(k)})$ under the current solution $u_h^{(k)}$:

$$(a_{i,j}^{(k)}, \theta_{i,j}^{(k)}) = \underset{(a_{i,j}, \theta_{i,j}) \in \Gamma_{i,j}}{\arg\max} \left\{ A_h(a_h, \theta_h) u_h^{(k)} - b_h(a_h, \theta_h) \right\}_{i,j}, \tag{27}$$

for all $\mathbf{x}_{i,j} \in \Omega$. Here $\Gamma_{i,j} = [0, 1] \times [-\frac{\pi}{4}, \frac{\pi}{4})$ is the control set at $\mathbf{x}_{i,j}$.

Meanwhile, obtain the residual

$$r_h^{(k)} = A_h(a_h^{(k)}, \theta_h^{(k)}) u_h^{(k)} - b_h(a_h^{(k)}, \theta_h^{(k)}). \tag{28}$$

2. If $\|r_h^{(k)}\| \leq$ tolerance: break

   Else, *use the multigrid V-cycle* to solve the following linear system for the solution $u_h^{(k+1)}$ under the current optimal control pair $(a_h^{(k)}, \theta_h^{(k)})$:

$$A_h(a_h^{(k)}, \theta_h^{(k)})\, u_h^{(k+1)} = b_h(a_h^{(k)}, \theta_h^{(k)}) \;\Rightarrow\; u_h^{(k+1)}. \tag{29}$$

To summarize, in order to solve (25), the inner multigrid V-cycle iteration for linearized problems is nested in an outer policy iteration. For convenience, we refer this type of multigrid methods as "policy-MG iteration".

The advantage of using this approach is that policy iteration is guaranteed to converge for any initial guess $u_h^{(0)}$, if HJB equation is monotonically discretized [3, 7]. Policy iteration consists of two sub-steps. The first sub-step is to solve the optimization problem at each grid point $\mathbf{x}_{i,j}$; see (27). Our recent work [10] discusses speeding up computation of the optimization problem in details. The second sub-step of the policy iteration is to solve the linear system under a given control pair; see (29). The second sub-step is our focus of developing multigrid methods.

## 4.2 *MG for 7-Point Stencil*

The components of the standard multigrid include pointwise smoother, full coarsening, full-weighting restriction, bilinear interpolation and coarse grid operator (i.e., Galerkin coarse grid operator or direct discretization). However, the standard multigrid leads to a poor convergence for the HJB equation. We need to adapt each multigrid component to the HJB equation in order to achieve fast convergence.

### 4.2.1 Nonlinear Smoother

First, we discuss smoothers. We observe that (11) may become anisotropic. For instance, if $c^* = \epsilon$ is a small constant close to 0 and $\theta^* = 0$, then (11) becomes

$$-\epsilon u_{xx} - (1-\epsilon)u_{yy} + 2\sqrt{\epsilon(1-\epsilon)f} = 0,$$

which is an anisotropic Poisson equation. It is well-known that when solving anisotropic equations, the standard pointwise smoothers do not smooth errors along the weakly connected axis, which causes poor convergence rates [42].

To address anisotropy, we consider using line smoothers. More specifically, instead of updating the unknowns point by point, we update strongly-connected grid points collectively. In general, the strongly-connected direction of the 7-point discretization can change alignment to either the $x$-axis, or the $y$-axis, or the diagonal axes, in different parts of the computational domain. In view of this, we apply

**four-direction alternating Gauss-Seidel line smoother**. Thus, the line smoother is applied four times: along the $x$-axis (left to right), the $y$-axis (top to bottom), the diagonal axis (top left to bottom right) and the transpose diagonal axis (top right to bottom left). We summarize the nonlinear smoother in Algorithm 2.

---

**Algorithm 2** Nonlinear four-direction alternating Gauss-Seidel line smoother

---

1: **subroutine** $\bar{u}_h$ = SMOOTH $(u_h)$
2: **for** $i = 1, , ..., n_x$ **do**
3: **for** $j = 1, , ..., n_y$ **do**
4: Update the control: $(\bar{c}_{i,j}, \bar{\theta}_{i,j}) = \underset{(c_{i,j},\theta_{i,j})\in\Gamma}{\arg\max}\ \mathcal{L}_{i,j}(c_{i,j}, \theta_{i,j}; u_h)$.
5: **end for**
6: **end for**
7: Apply the one-step four-direction alternating Gauss-Seidel line smoother to the linearized system $A_h(\bar{c}_h, \bar{\theta}_h)\, u_h = b_h(\bar{c}_h, \bar{\theta}_h)$, which updates the solution $u_h \to \bar{u}_h$.

---

### 4.2.2 Restriction and Interpolation

Once the error becomes smooth along the $x$, $y$ and diagonal axes after using the four-direction alternating line smoother, the standard full-coarsening can be applied. In order to capture the directional feature of the 7-point discretization, we follow [42] and apply 7-point restriction operators to (19). Using the stencil notation introduced, the corresponding 7-point restriction operators are given by

$$R^{[1]} = \frac{1}{8}\begin{bmatrix} 0 & 1 & 1 \\ 1 & 2 & 1 \\ 1 & 1 & 0 \end{bmatrix}, \quad R^{[2]} = \frac{1}{8}\begin{bmatrix} 1 & 1 & 0 \\ 1 & 2 & 1 \\ 0 & 1 & 1 \end{bmatrix}, \tag{30}$$

respectively. The interpolation operator is the scaled transpose of the restriction operator:

$$P = 4R^T. \tag{31}$$

## 4.3 *MG for Mixed Discretization*

In this section, we will discuss multigrid methods for the more general mixed discretization, where the semi-Lagrangian wide stencil discretization is applied to part of the computational domain. We will propose global linearization multigrid methods instead of FAS methods. One reason is that mixed discretization with wide stencils is a more difficult problem than the pure standard 7-point stencil discretization. We would like to use the Petrov-Galerkin coarse grid operators, which is more robust in terms of the accuracy of the error estimate but is incompatible with the nonlinearity of FAS. Another reason, which will be shown, is that the coarse grids of our proposed approach are no longer square grids, which poses difficulties in defining an FAS coarse grid problem using direct discretization.

### 4.3.1 Issues

To start with a simple scenario, we consider solving the mixed discretization of the following linearized HJB equation:

$$\begin{aligned} \frac{1}{2}u_{xx} + \frac{1}{2}u_{yy} &= \sqrt{f}, \quad \text{in } \Omega\backslash\{(0,0)\}, \\ \frac{2+\sqrt{2}}{4}u_{xx} + \frac{2-\sqrt{2}}{4}u_{yy} + \frac{1}{\sqrt{2}}u_{xy} &= 0, \quad \text{at } (0,0), \\ u &= g, \quad \text{on } \partial\Omega. \end{aligned} \tag{32}$$

In other words, we assume that the control is given as $(c^*, \theta^*) = (\frac{1}{2}, 0)$ on the entire computational domain $\Omega$, where the standard 7-point stencil discretization is applied, except that the control is $(c^*, \theta^*) = (1, \frac{\pi}{8})$ at the origin (the center of $\Omega$), where wide stencil discretization is applied. Figure 3(ii) shows the error after applying the four-direction alternating line smoother. In particular, the cross section of the smoothed error shows that a kink appears at the origin (0,0). In general, wherever the wide stencil discretization is applied at a grid point, a kink appears in a smoothed error. Unfortunately, such kinks cannot be eliminated by other types of smoothers either.

### 4.3.2 Coarsening Strategy

Despite kink(s), Fig. 3(ii) shows that, after smoothing, kink(s) are restricted to the wide stencil point(s), and the error at the other grid points (i.e., the standard 7-point stencil points) is still smooth. This motivates us to apply full-coarsening to the standard 7-point stencil points, and consider a special type of coarsening strategy at the wide stencil points.

To motivate our coarsening strategy for wide stencils, we define a $C$-point as a fine grid point that is kept in its corresponding coarse grid; and an $F$-point otherwise. Let us first consider a one-dimensional cross section of a smoothed error; see Fig. 4(i). Black dots are $C$-points, while hollow dots are $F$-points. Assume that the standard full-coarsening assigns a wide stencil point (indicated by the red arrow) as an $F$-point. Let the black curves represent the underlying fine grid error. On the coarse grid, let its estimated error match the underlying fine grid error exactly, i.e., let the values of the black dots sit on the black curve. After linear interpolation of the coarse grid error, we obtain the interpolated error (grey curve) on the fine grid. Ideally, the interpolated error (grey curve) should match the underlying fine grid error (black

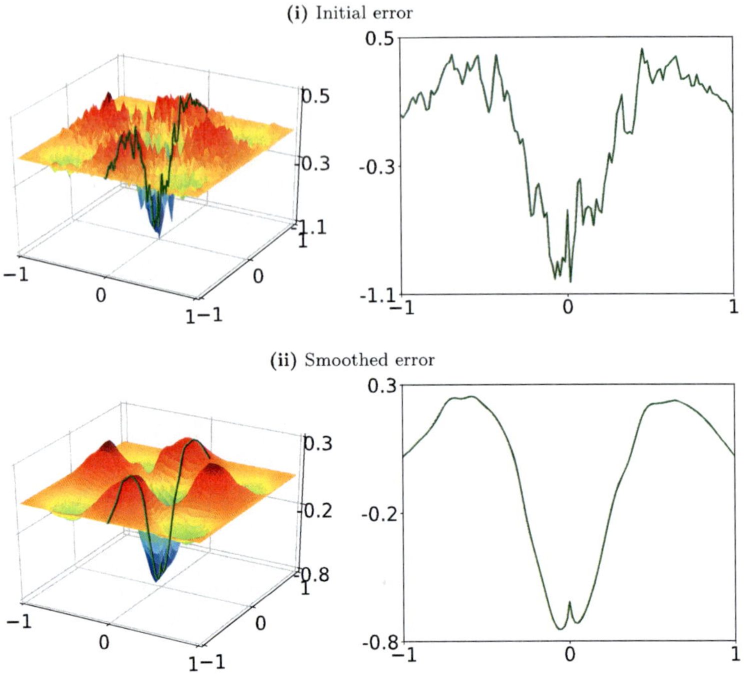


**Fig. 3** The error after one step four-direction alternating Gauss-Seidel line smoothing. **(i)** Initial error and its cross section along the $x$-axis. **(ii)** Smoothed error and its cross section along the $x$-axis. A kink appears at the origin (0,0)

curve) as closely as possible. However, since the underlying fine grid error has a kink at the wide stencil point, the resulting interpolated error turns out to have a mismatch, as indicated by the red arrow. In other words, if the wide stencil point is an $F$-point, a linearly interpolated error will fail to capture the kink accurately.

Instead, our approach is simply setting the wide stencil $F$-point as a coarse grid point, i.e., a $C$-point; see Fig. 4(ii). As a result, interpolation at the wide stencil point is no longer needed. The error at the wide stencil point is simply copied from the coarse grid to the fine grid. This yields a more accurate fine grid estimated error, as indicated by the green arrow.

The above coarsening strategy can be extended to two dimensions. Figure 5 illustrates the coarsening process. On the fine grid, the black dots are selected as $C$-points, and the hollow dots are selected as $F$-points. Suppose wide stencils are applied to the three red dots. Then these three dots are all assigned as $C$-points. The resulting first coarse grid is a combination of a square grid that comes from geometric coarsening, and some additional coarse grid points that come from wide stencils. We can continue to coarsen the square sub-grid and meanwhile keep all the wide stencil points as $C$-points, which generates the second coarse grid. Such a coarsening strategy can be applied recursively until the coarsest level.

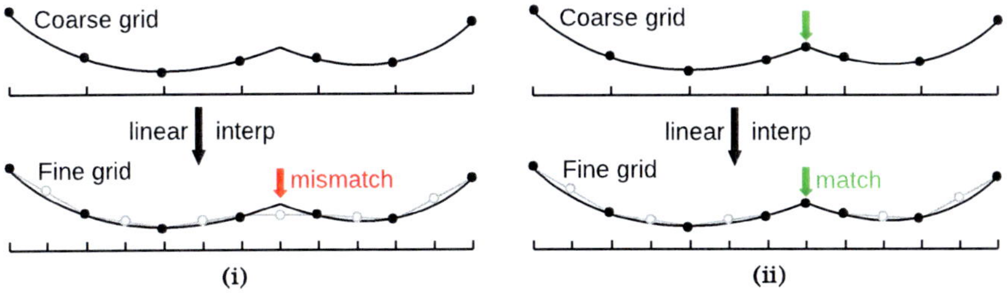


**Fig. 4** Coarsening strategy at a wide stencil point. **(i)** Standard coarsening with linear interpolation at a wide stencil $F$-point (red arrow). **(ii)** Setting the wide stencil point as a coarse grid $C$-point (green arrow)

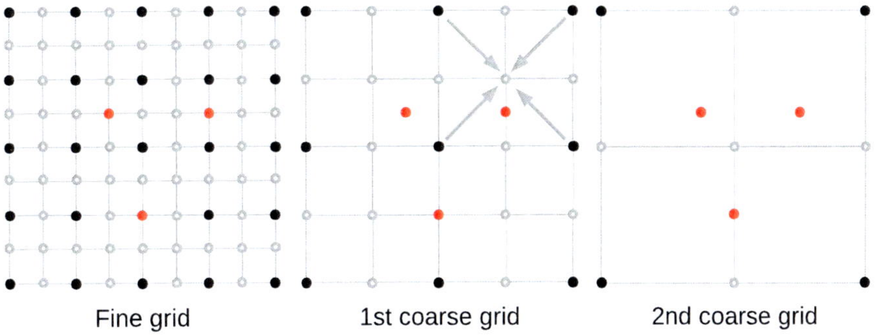


**Fig. 5** Wide stencil grid points (red) are kept as $C$-points as the grid is coarsened from a fine grid to a coarse grid

One may argue that by setting all the wide stencil points as coarse grid points, the number of coarse grid points, and thus the computational complexity, will increase. However, it is observed in numerical simulations that wide stencils typically account for a negligible proportion of the total grid points in practical applications (such as image registration). Setting wide stencil points as coarse grid points would not result in a significant increase of the number of coarse grid points, and would still approximately maintain the square grid structure as the grid coarsens.

### 4.3.3 Interpolation

Under the proposed coarsening strategy, all the wide stencil points are excluded from the set of $F$-points. In other words, $F$-points must be the standard 7-point stencils. Hence, the 7-point interpolation, as described in Sect. 4.2.2, can be used for interpolating the errors at these $F$-points.

We note that the coarse grids are no longer square grids; see Fig. 5. However, each of these coarse grids can be seen as a combination of a square grid and some additional wide-stencil $C$-points. Then all the $F$-points can still be interpolated from

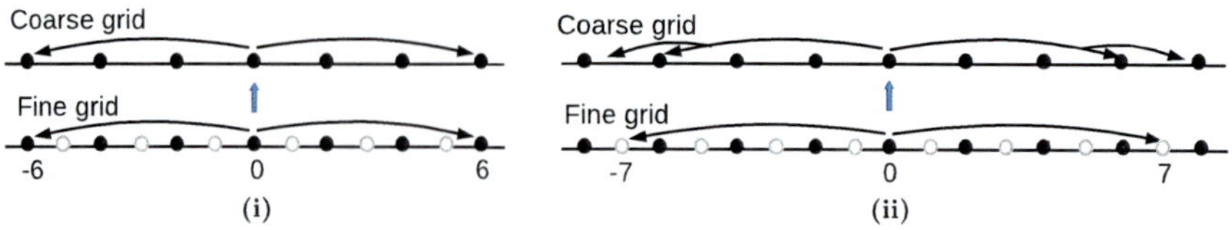


**Fig. 6** Restriction for one-dimensional Poisson equation. **(i)** $h = \frac{1}{36}$ and $\sqrt{h} = 6h$. **(ii)** $h = \frac{1}{49}$ and $\sqrt{h} = 7h$

the $C$-points on the square grid. The arrows in Fig. 5 show how an $F$-point can be interpolated.

### 4.3.4 Restriction

In both the standard geometric and algebraic multigrid methods, restriction is simply the transpose of interpolation. However, it does not result in mesh-independent convergence rates for the non-symmetric matrices $A_h$ arising from the mixed discretization. We will show such poor convergence in Sect. 5.2. Instead, we propose a restriction operator $R$ that is different from the transpose of the interpolation $P$.

Our approach is simply to use **injection** on wide stencil points. To motivate the use of injection, let us simplify our problem and start with the one-dimensional Poisson equation

$$-u_{xx} = 0, \quad x \in [-0.5, 0.5]. \tag{33}$$

We apply the wide stencil discretization at $x = 0$ and the standard finite difference discretization on the rest of the computational domain. Figure 6 shows that under our coarsening strategy (which in this case is the same as the standard full coarsening), the fine grid points with even indices are $C$-points (black points), and the ones with odd indices are $F$-points (hollow points). The wide stencil point is $i = 0$. A naive choice of restriction at $i = 0$ would be the transpose of the linear interpolation, i.e., the standard full-weighting restriction:

$$r_0^H = \frac{1}{4} r_{-1} + \frac{1}{2} r_0 + \frac{1}{4} r_1, \tag{34}$$

where $r_{-1}$, $r_0$, $r_1$ are the fine grid residuals at $i = -1, 0, 1$, respectively, and $r_0^H$ is the restricted residual at the coarse grid point. However, this leads to a poor coarse grid estimated error. In order to find a better restriction, we investigate two cases.

**Case 1:** $h = \frac{1}{36}$ and $\sqrt{h} = 6h$. Figure 6(i) shows that on the fine grid, the stencil points of $i = 0$ fall onto $i = \pm 6$. In this case, the wide stencil discretization at $i = 0$ reads

$$\frac{-u_{-6} + 2u_0 - u_6}{(6h)^2} = 0. \tag{35}$$

The residual at $i = 0$ is then given by

$$r_0 = \frac{-e_{-6} + 2e_0 - e_6}{(6h)^2}. \tag{36}$$

We notice that $i = 0, i = -6$ and $i = 6$ are all $C$-points. Then a natural construction of the coarse grid problem at $i = 0$ is to discretized the Poisson equation using these three points, or more precisely,

$$\frac{-e_{-6}^H + 2e_0^H - e_6^H}{(6h)^2} = r_0^H, \tag{37}$$

where the left hand side is a discretization of the Poisson equation on the coarse grid with the stencil length $6h$, and the right hand side is the coarse grid residual $r_0^H$. Comparing (36) and (37), we can see that the restriction at $i = 0$ is a simple injection:

$$r_0^H \equiv r_0. \tag{38}$$

**Case 2:** $h = \frac{1}{49}$ and $\sqrt{h} = 7h$. Figure 6(ii) shows that on the fine grid, the stencil points of $i = 0$ fall onto $i = \pm 7$. Unlike the previous case, here the two points $i = \pm 7$ are both $F$-points. To discretize the Poisson equation on the coarse grid, we interpolate the errors at $i = 7$ and $i = -7$ from their neighboring $C$-points, which gives

$$\frac{-\frac{1}{2}(e_{-8}^H + e_{-6}^H) + 2e_0^H - \frac{1}{2}(e_6^H + e_8^H)}{(7h)^2} = r_0^H. \tag{39}$$

We want to find a restriction, i.e., to rewrite $r_0^H$ as a linear combination of fine grid residuals, such that it matches the left hand side of (39). One scheme is to use the linear combination of the following fine grid residuals:

$$r_0 = \frac{-e_{-7} + 2e_0 - e_7}{(7h)^2}, \quad r_7 = \frac{-e_6 + 2e_7 - e_8}{h^2}, \quad r_{-7} = \frac{-e_{-6} + 2e_{-7} - e_{-8}}{h^2}. \tag{40}$$

If we combine $r_0$, $r_7$ and $r_{-7}$ as follows

$$r_0 + \frac{1}{98}r_7 + \frac{1}{98}r_{-7} = \frac{-\frac{1}{2}(e_{-8} + e_{-6}) + 2e_0 - \frac{1}{2}(e_6 + e_8)}{(7h)^2}, \tag{41}$$

then (41) matches the left hand side of (39) in the exact sense. Equation (41) defines a possible restriction, i.e.,

$$r_0^H \equiv r_0 + \frac{1}{98}r_7 + \frac{1}{98}r_{-7}. \tag{42}$$

We note that the restriction (41) makes use of the residuals $r_7$ and $r_{-7}$, which are the points that the wide stencil point $i = 0$ connects to. This is different from the standard full weighting restriction (34), which uses the neighboring points $r_1$ and $r_{-1}$. Since the coefficients of $r_7$ and $r_{-7}$ are small, we simply drop them from (42) and yield again an injection:

$$r_0^H \equiv r_0. \tag{43}$$

More generally, given a wide stencil $C$-point $i \in C$ with a stencil length $\sqrt{h}$, the non-zero restriction weights occur at the set of the $F$-points that it connects to, denoted as $\{j \mid j \in F, A_{i,j} \neq 0\}$. We can show that the restriction weights are

$$w_{i,j} = -\frac{A_{i,j}}{A_{j,j}} = -\frac{-\frac{1}{(\sqrt{h})^2}}{\frac{2}{h^2}} = \frac{h}{2}. \tag{44}$$

When $h$ is small, the restriction (44) can be left out. In other words, injection is sufficient for a good coarse grid problem.

We extend the proposed injection at wide stencil $C$-points from the one-dimensional Poisson equation to the two-dimensional HJB equation. Note that the resulting restriction operator $R_h$ is no longer the transpose of the interpolation. Once the restriction operator is specified, we construct the coarse grid operator by

$$A_{2h} \equiv R_h A_h P_h. \tag{45}$$

Since $R_h \neq P_h^T$, it results in the Petrov-Galerkin coarse grid operator.

The benefits of injection at wide stencil $C$-points are two-fold. One is that the resulting restriction operator and Petrov-Galerkin operator (45) are significantly sparser than their counterparts if other types of restriction operators are used (such as AMG restriction). This reduces the computational complexity. The other benefit is that such restriction would lead to an accurate coarse grid error estimate and eventually a mesh-independent convergence rate (Fig. 7).

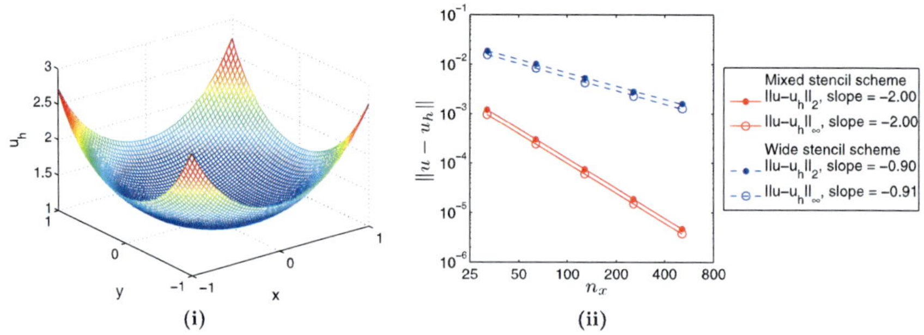


**Fig. 7** Example 1: The exact solution is $u(x, y) = e^{\frac{1}{2}(x^2+y^2)}$. **(i)** Numerical solution. **(ii)** Norms of the errors $\|u - u_h\|$

## 5 Numerical Results

In this section, we demonstrate the mesh-independent convergence rates of the proposed multigrid methods for solving the discretized system (25)–(26). Details can be found in [10, 11].

### 5.1 *Multigrid for Standard 7-Point Stencil Discretization*

In Examples 1–2, the standard 7-point stencil discretization can be applied monotonically on the entire computational domain. We compare the performance of two families of multigrid methods - global linearization methods and full approximation scheme (FAS). For global linearization methods, the residual tolerances for the outer policy iteration and the inner multigrid V-cycle are $10^{-6}$ and $10^{-7}$, respectively. The Gauss-Seidel smoother, the standard full coarsening and the 7-point restriction and interpolation are applied. The Petrov-Galerkin coarse grid operators are used to construct coarse grid problems. For FAS, the multigrid components are the same as the global linearization methods, except that we use the nonlinear version of the smoothers and direct discretization coarse grid operators.

**Example 1** Consider solving the following equation:

$$\begin{aligned} u_{xx}u_{yy} - u_{xy}^2 &= f(x,y) = (1 + x^2 + y^2)e^{x^2+y^2}, \qquad && \text{in } \Omega, \\ u(x,y) &= g(x,y) = e^{\frac{1}{2}(x^2+y^2)}, && \text{on } \partial\Omega, \end{aligned}$$

where $\Omega = (-1, 1) \times (-1, 1)$. The exact solution $u(x, y) = e^{\frac{1}{2}(x^2+y^2)}$ is smooth.

This example is isotropic, so it suffices to apply the less expensive pointwise Gauss-Seidel smoother. First we show the convergence rates of the global linearization method; see the first and second columns of Table 1. To understand the reported numbers, we take the grid size of $32 \times 32$ as an example. The numbers "8, 7, 2" mean that it takes 3 policy iterations to converge to the solution of the nonlinear

**Table 1** Convergence of the global linearization method and the FAS for Example 1

| $n_x \times n_y$ | Gobal linearization method | | FAS |
|---|---|---|---|
| | Number of multigrid V-cycles within each policy iteration | Total number of multigrid V-cycles | Total number of multigrid V-cycles |
| $32 \times 32$ | 8, 7, 2 | 17 | 8 |
| $64 \times 64$ | 9, 7, 3 | 19 | 8 |
| $128 \times 128$ | 9, 7, 3 | 19 | 9 |
| $256 \times 256$ | 9, 7, 3 | 19 | 9 |

problem, where the 1st policy iteration takes 8 V-cycles to converge to the solution of the linearized problem, the 2nd policy iteration takes 7 V-cycles, and the 3rd policy iteration takes 2 V-cycles. The table shows that the number of multigrid V-cycles within each policy iteration ranges from 2–9. The total number of multigrid V-cycles for solving the nonlinear problem is 17–19, independent of mesh size. As a side remark, we use the solution of the $k$-th policy iteration, $u_h^{(k)}$, as the initial guess of the multigrid V-cycles at the $(k+1)$-th policy iteration. Hence, as policy iteration converges, the initial guess of multigrid V-cycles becomes more and more precise, and the number of multigrid V-cycles within each policy iteration decreases.

We compare the global linearization method with the FAS iteration. The last column of Table 1 shows that the total number of the FAS iterations is 8–9 and is independent of mesh size. We note that for both the global linearization method and the FAS iteration, the computational cost per multigrid iteration is approximately the same. Hence, the FAS iteration is less expensive and converges faster.

**Example 2** We consider the following equation:

$$\begin{aligned} u_{xx}u_{yy} - u_{xy}^2 &= f(x, y) = 1 + 24(x + y)^2, && \text{in } \Omega, \\ u(x, y) &= g(x, y) = \tfrac{1}{2}(x^2 + y^2) + (x + y)^4, && \text{on } \partial\Omega. \end{aligned}$$

The exact solution is $u(x, y) = \frac{1}{2}(x^2 + y^2) + (x + y)^4$.

Table 2 reports the convergence of the global linearization method using alternating line smoother and pointwise smoother. The multigrid V-cycle with the alternating line smoother converges at 20–32 iterations in total, which is approximately independent of mesh size. Conversely, the multigrid V-cycle with a pointwise smoother converges with more than 70 iterations, and the number of iterations is more than doubled as $n_x$ increases from 32 to 256. This is because the example is anisotropic, and a pointwise smoother is not efficient in smoothing errors along weakly connected directions.

Similar to Example 1, we also compare the total numbers of multigrid V-cycles given by the global linearization method with the numbers given by the FAS. The

**Table 2** Convergence of the global linearization method for Example 2 using alternating line smoother and pointwise smoother

| $n_x \times n_y$ | MG with alternating line smoother | | MG with pointwise smoother |
|---|---|---|---|
| | Number of multigrid V-cycles within each policy iteration | Total number of multigrid V-cycles | Total number of multigrid V-cycles |
| $32 \times 32$ | 5,5,5,3,2 | 20 | 73 |
| $64 \times 64$ | 5,6,6,4,2,1 | 24 | 94 |
| $128 \times 128$ | 6,6,7,5,3,1 | 28 | 129 |
| $256 \times 256$ | 7,7,7,6,3,1 | 32 | 161 |

**Table 3** Total number of multigrid V-cycles of the global linearization method and the FAS for Example 2 using the alternating line smoother

| $n_x \times n_y$ | Global linearization method | FAS |
|---|---|---|
| 32 × 32 | 20 | 5 |
| 64 × 64 | 24 | 6 |
| 128 × 128 | 28 | 6 |
| 256 × 256 | 32 | 6 |

alternating line smoother is used. Table 3 shows that the global linearization method converges in 20–32 iterations, whereas the FAS converges in 5–6 iterations, which is significantly faster.

## 5.2 *Multigrid for Mixed Discretization*

In this section, we illustrate the multigrid convergence rates for the mixed discretization. Thus, we apply four-direction alternating line smoother. At standard 7-point stencil points, we apply the standard full coarsening and the 7-point restriction and interpolation. At wide stencil points, we set them as coarse grid points, and use injection as the restriction. The Petrov-Galerkin coarse grid operators are used for constructing coarse grid problems.

**Example 3** We consider solving the linearized HJB equation (32), where $f$ and $g$ are the same as in Example 1. Consider applying the wide stencil at the origin and the standard 5-point stencil discretization everywhere else. We compare the performance of our multigrid method (Scheme I), the standard multigrid with four-direction alternating line smoother (Scheme II), and the standard multigrid with pointwise Gauss-Seidel smoother (Scheme III). For this example, the only difference between Schemes I and II is that injection is applied at the wide stencil point for Scheme I, while full-weighting restriction is applied at the same point for Scheme II. Table 4 shows that Scheme III has poor convergence. Scheme II converges in less than 20 iterations, but the convergence rate grows as $n_x$ increases. Scheme I converges in 5–6 iterations, and the convergence rate is independent of mesh size.

Figure 8 explains the convergence observed in Table 4 by examining the evolution of errors during one two-grid cycle. Only the cross sections along the x-axis are plotted. Start with the same initial error (green lines) for both our and the standard schemes. The pre-smoothed error (blue lines) is smooth everywhere, except that a kink appears at the wide stencil point $x = 0$. Figure 8(i) uses our algorithm, where injection is applied at the wide stencil point $x = 0$. The resulting coarse grid problem yields an accurate coarse grid estimated error, i.e., the red line matches the blue line well. Such accurate coarse grid estimate eliminates the error effectively, and yields

**Table 4** Convergence of linear multigrid V-cycles for Example 3

| $n_x \times n_y$ | **Scheme I: Our MG** | Scheme II: Standard MG with alternating line smoother | Scheme III: Standard MG with pointwise smoother |
|---|---|---|---|
| $32 \times 32$ | **5** | 7 | 23 |
| $64 \times 64$ | **5** | 9 | 46 |
| $128 \times 128$ | **6** | 12 | 198 |
| $256 \times 256$ | **6** | 17 | more than 200 |

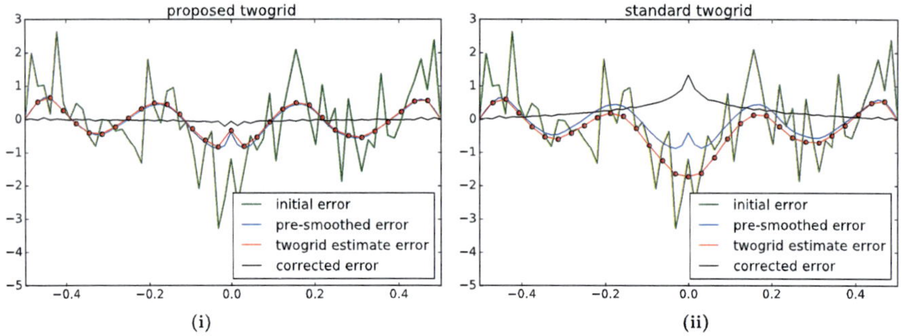


**Fig. 8** Cross sections of errors along the x-axis. **(i)** Our algorithm, where injection is used at the wide stencil point $x = 0$. **(ii)** Standard algorithm, where full-weighting restriction is used

a small post-corrected error (black line). Conversely, under the same smoother, if the standard full-weighting is used at the wide stencil, then Fig. 8(ii) shows that the coarse grid estimated error (red line) is no longer a good approximation of the pre-smoothed error (blue line).

**Example 4** We use the global linearization method to solve the Monge-Ampère equation as in Example 1, where

$$f(x, y) = \max\left(1 - \frac{0.15}{\sqrt{x^2 + y^2}}, 0\right), \quad g(x, y) = \frac{1}{2}(\sqrt{x^2 + y^2} - 0.15)^2$$

on $\Omega = (-0.5, 0.5) \times (-0.5, 0.5)$. The viscosity solution is given by $u(x, y) = \frac{1}{2}\max(\sqrt{x^2 + y^2} - 0.15, 0)^2$. This is a $C^1$ function where the solution is not smooth at the ring $x^2 + y^2 = 0.15^2$. Semi-Lagrangian wide stencils are applied near the ring (Fig. 9).

Table 5 reports the convergence of the global linearization method. The number of outer policy iterations increases from 5 to 10 as $n_x$ increases from 32 to 256. Such increase of outer iteration is related to nonlinearity and the singularity on the ring.

To compare the number of multigrid V-cycles across different mesh sizes fairly, we compute the average number of multigrid V-cycles per policy iteration. Table 5

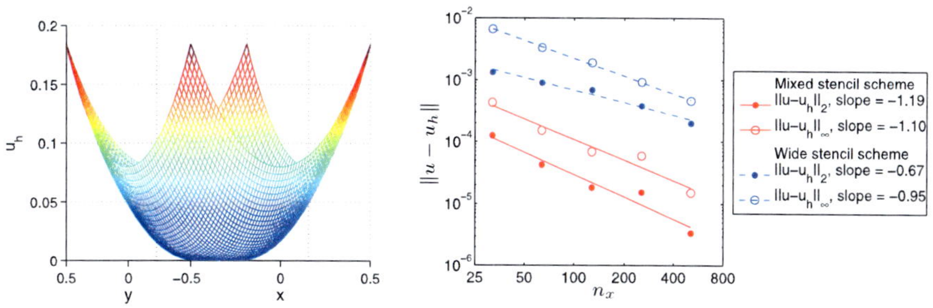


(i) (ii)

**Fig. 9** Example 3: The exact solution is $\frac{1}{2}\max(\sqrt{x^2+y^2} - 0.15, 0)^2$. **(i)** Numerical solution. **(ii)** Norms of the error $\|u - u_h\|$

**Table 5** Convergence of the global linearization multigrid method for Example 4

| $n_x \times n_y$ | Number of multigrid V-cycles within each policy iteration | Average number of multigrid V-cycles per policy iteration |
|---|---|---|
| $32 \times 32$ | 4,5,3,2,1 | 3.0 |
| $64 \times 64$ | 4,6,3,2,1 | 3.2 |
| $128 \times 128$ | 5,6,4,3,3,2 | 3.8 |
| $256 \times 256$ | 6,6,6,6,5,4,3,3,2,1 | 4.2 |

shows that the average V-cycle count is approximately a constant ranging from 3.0 to 4.2 as $n_x$ increases from 32 to 256. Hence, the inner multigrid V-cycle for solving linearized systems is nearly mesh-independent.

## 6 Conclusion

This paper presents a numerical scheme for solving the mass transport registration model. In particular, we introduce a mixed standard 7-point stencil and wide stencil finite difference discretization. Furthermore, we present multigrid methods for solving the mixed discretization of the Monge-Ampère equation. We investigate two scenarios. One scenario is when the standard 7-point stencil discretization is applied on the entire computational domain. FAS gives the optimal mesh-independent convergence. The other scenario is the general mixed discretization. Global linearization method is used. We set all wide stencil points as coarse grid points and propose injection of residuals at wide stencil points. The resulting multigrid methods converge at mesh-independent rates.

## References


1. J.B. Antoine Maintz, M.A. Viergever, A survey of medical image registration. Medical Image Anal. **2**(1), 1–36 (1998)
2. A. Ardeshir Goshtasby, *2-D and 3-D Image Registration: For Medical, Remote Sensing, and Industrial Applications* (Wiley, 2005)
3. P. Azimzadeh, P.A. Forsyth, Weakly Chained Matrices, Policy Iteration, and Impulse Control. SIAM J. Numer. Anal. **54**(3), 1341–1364 (2016)
4. G. Barles, P.E. Souganidis, Convergence of approximation schemes for fully nonlinear second order equations. Asymptotic Anal. **4**(3), 271–283 (1991)
5. J.-D. Benamou, Y. Brenier, K. Guittet, The Monge-Kantorovitch mass transfer and its computational fluid mechanics formulation. Int. J. Numer. Methods Fluids **40**(1–2), 21–30 (2002). ICFD Conference on Numerical Methods for Fluid Dynamics (Oxford, 2001)
6. J.-D. Benamou, B.D. Froese, A.M. Oberman, Two numerical methods for the elliptic Monge-Ampère equation. M2AN Math. Model. Numer. Anal. **44**(4), 737–758 (2010)
7. Olivier Bokanowski, Stefania Maroso, and Hasnaa Zidani. Some convergence results for Howard's algorithm. SIAM J. Numer. Anal., 47(4):3001–3026, 2009
8. C. Broit, *Optimal Registration of Deformed Images* (1981)
9. Rick Chartrand, Brendt Wohlberg, Kevin R. Vixie, and Erik M. Bollt. A gradient descent solution to the Monge-Kantorovich problem. Appl. Math. Sci. (Ruse), 3(21–24):1071–1080, 2009
10. Y. Chen, J.W.L. Wan, J. Lin, Monotone mixed finite difference scheme for Monge–Ampère equation. J. Sci. Comput. **76**, 1839–1867 (2018)
11. Y. Chen, J.W.L. Wan, Multigrid methods for convergent mixed finite difference scheme for Monge-Ampère equation. Comput. Visual. Sci., 1–15 (2017)
12. Y. Chen, J.W.L. Wan, Numerical method for image registration model based on optimal mass transport. Inverse Probl. Imaging **12**(2), 401–432 (2018)
13. S.N. Chow, W. Li, H. Zhou, A discrete Schrödinger equation via optimal transport on graphs. Journal of Functional Analysis **276**, 2440–2469 (2019)
14. G.E. Christensen, *Deformable shape models for anatomy*. Ph.D. thesis, Washington University Saint Louis, Mississippi (1994)
15. Kristian Debrabant and Espen R. Jakobsen. Semi-Lagrangian schemes for linear and fully non-linear diffusion equations. Math. Comp., 82(283):1433–1462, 2013
16. P. Dupuis, U. Grenander, M.I. Miller, Variational problems on flows of diffeomorphisms for image matching. Quart. Appl. Mathe., 587–600 (1998)
17. Xiaobing Feng, Roland Glowinski, and Michael Neilan. Recent developments in numerical methods for fully nonlinear second order partial differential equations. SIAM Rev., 55(2):205–267, 2013
18. B. Fischer, J. Modersitzki, Fast inversion of matrices arising in image processing. Numerical Algorithms **22**(1), 1–11 (1999)
19. P.A. Forsyth, G. Labahn, Numerical methods for controlled Hamilton-Jacobi-Bellman PDEs in finance. Journal of Computational Finance **11**(2), 1 (2007)
20. Brittany D. Froese. A numerical method for the elliptic Monge-Ampère equation with transport boundary conditions. SIAM J. Sci. Comput., 34(3):A1432–A1459, 2012
21. Brittany D. Froese and Adam M. Oberman. Convergent finite difference solvers for viscosity solutions of the elliptic Monge-Ampère equation in dimensions two and higher. SIAM J. Numer. Anal., 49(4):1692–1714, 2011
22. S. Haker, A. Tannenbaum, Optimal mass transport and image registration, in *Variational and Level Set Methods in Computer Vision, 2001. Proceedings. IEEE Workshop on*, pp. 29–36. IEEE (2001)
23. S. Haker, L. Zhu, A. Tannenbaum, S. Angenent, Optimal mass transport for registration and warping. International Journal of computer vision **60**(3), 225–240 (2004)
24. D.L.G. Hill, P.G. Batchelor, M. Holden, D.J. Hawkes, Medical image registration. Phys. Medicine Biol. **46**(3), R1 (2001)

25. R.A. Howard, *Dynamic Programming and Markov Processes*. The Technology Press of M.I.T., Cambridge, Mass (Wiley, New York-London, 1960)
26. M. Irani, S. Peleg, Improving resolution by image registration. CVGIP: Graph. Models Image Process. **53**(3), 231–239 (1991)
27. M. Knott, C.S. Smith, On the optimal mapping of distributions. J. Optim. Theory Appl. **43**(1), 39–49 (1984)
28. N.V. Krylov, The control of the solution of a stochastic integral equation. Teor. Verojatnost. i Primenen. **17**, 111–128 (1972)
29. P.-L. Lions, Hamilton-Jacobi-Bellman equations and the optimal control of stochastic systems, in *Proceedings of the International Congress of Mathematicians, Vol. 1, 2 (Warsaw, 1983)*, pp. 1403–1417. PWN, Warsaw (1984)
30. Lisa Gottesfeld Brown, A survey of image registration techniques. ACM Comput. Surv. (CSUR) **24**(4), 325–376 (1992)
31. K. Ma, P.A. Forsyth, An unconditionally monotone numerical scheme for the two factor uncertain volatility model. Preprint (2014)
32. J. Modersitzki, *Numerical Methods for Image Registration*. Numerical Mathematics and Scientific Computation. Oxford University Press, New York, 2004. Oxford Science Publications
33. Oleg Museyko, Michael Stiglmayr, Kathrin Klamroth, and Günter Leugering. On the application of the Monge-Kantorovich problem to image registration. SIAM J. Imaging Sci., 2(4):1068–1097, 2009
34. G. Peyré, M. Cuturi, Computational optimal transport. Foundations and Trends in Machine Learning **51**(1), 1–44 (2019)
35. C. Reisinger, J.R. Arto, Boundary treatment and multigrid preconditioning for semi-Lagrangian schemes applied to Hamilton-Jacobi-Bellman equations. arXiv preprint arXiv:1605.04821 (2016)
36. K. Rohr, *Landmark-Based Image Analysis: Using Geometric and Intensity Models*, vol. 21 (Springer Science & Business Media, 2001)
37. Louis-Philippe Saumier, Martial Agueh, and Boualem Khouider. An efficient numerical algorithm for the $L^2$ optimal transport problem with periodic densities. IMA J. Appl. Math., 80(1):135–157, 2015
38. M. Schmitz, M. Heitz, N. Bonneel, F.M. Ngole Mboula, D. Coeurjolly, M. Cuturi, G. Peyré, J-L. Starck, Wasserstein dictionary learning: Optimal transport-based unsupervised non-linear dictionary learning. SIAM J. Imaging Sci. **11**(1), 643–678 (2018)
39. Benjamin Seibold. Performance of algebraic multigrid methods for non-symmetric matrices arising in particle methods. Numer. Linear Algebra Appl., 17(2–3):433–451, 2010
40. A. Sotiras, C. Davatzikos, N. Paragios, Deformable medical image registration: A survey. IEEE transactions on medical imaging **32**(7), 1153–1190 (2013)
41. J.-P. Thirion, Image matching as a diffusion process: an analogy with maxwell's demons. Medical image analysis **2**(3), 243–260 (1998)
42. U. Trottenberg, C.W. Oosterlee, A. Schüller., textitMultigrid. Academic Press, Inc., San Diego, CA, 2001. With contributions by A. Brandt, P. Oswald and K. Stüben
43. A. Trouvé, Diffeomorphisms groups and pattern matching in image analysis. International Journal of Computer Vision **28**(3), 213–221 (1998)
44. P. Viola, W.M. Wells III, Alignment by maximization of mutual information. Int. J. Comput. Vis. **24**(2), 137–154 (1997)

# Author Index





X.-C. Tai et al. (eds.), *Mathematical Methods in Image Processing and Inverse Problems*,
Springer Proceedings in Mathematics & Statistics 360,
https://doi.org/10.1007/978-981-16-2701-9